CW01272331

The Lusitania Controversies
by Gary Gentile

On May 15, 1915, the British liner *Lusitania* fell victim to a German torpedo off the south coast of Ireland. In scarcely twenty minutes the crack Cunard steamship sank in 300 feet of water, leaving on the surface of the troubled green sea a blight of human flotsam: a handful of lifeboats, the cries of drowning passengers and crew, and the bodies of the dead. Among the 1,198 fatalities were scores of children and babes in arms, and 123 American citizens.

Death struck some with merciful swiftness. Most, however, languished in the frigid water and suffered the terrible numbing cold until their strength and endurance gave way. The silence that followed was more telling of the tragedy than the direful screams of the wretched.

The sinking of the *Lusitania* was the latest in a list of German atrocities that shocked the civilized world.

Shock quickly yielded to heated controversy as investigations and inquiries sought suitable answers to a host of provocative questions about the legality of unrestricted submarine warfare and the torpedoing of merchant vessels carrying none but innocent civilians. Was the *Lusitania* armed? Was she carrying a cargo of high explosives? Had her captain disregarded recent Admiralty warnings about U-boats operating in the vicinity? Did a minor infraction of international law make the *Lusitania* a legitimate target? And, philosophically, can an inhuman act be condoned or justified within the context of all-out war?

The Lusitania Controversies takes a bold and enlightened look at these and other issues that are still being argued today.

But there is more. The story then becomes an exciting tale of adventure: an impious trek through the history of deep wreck-diving from its meager beginnings in the 1950's, through miscarried commercial salvage operations which sought to wrest from the *Lusitania* the secrets that some people believed to exist, to the highly acclaimed mixed-gas diving expedition in 1994, in which the author took an active part.

The author brings original insight to this historical retelling because he has been deeply involved in deep wreck-diving and shipwreck research and writing for more than twenty-five years. He is considered a pioneer in the field of technical diving, and - through workshops and books - was instrumental in introducing mixed-gas diving technology to the realm of shipwreck exploration.

A saga of such incredible breadth and dimension could not be fitted between the covers of a single volume without excluding particulars that are compelling in the extreme. In this chronological account, Book One begins with the *Lusitania's* construction in 1905; covers her career, sinking, aftermath, and global repercussions; examines the liner's lasting impressions and publication history; discusses aborted and ineffective attempts to dive the wreck; then details the commencement and evolution of deep wreck-diving as far as 1979.

The untold story of early wreck-diving is rich in passion, portrayal, betrayal, and adventure. The author relates the exploits of his predecessors through unproven times when equipment and wreck exploration techniques were primitive. He follows the evolution of the activity through the slow maturation process until a greater understanding of the special methods and hazards was wrought after years of trial and effort.

Book Two picks up where Book One leaves off, carrying the story of deep wreck-diving forward from 1980. The second volume recounts a host of exciting explorations into the dark foreboding interior of the Italian liner *Andrea Doria*, which came to be a proving ground for deep wreck-divers the world over. Other exciting episodes abound as the author pursues ever-deeper challenges in the search for sunken shipwrecks.

Then came the application of helium breathing mixtures for scuba, which permitted divers to explore shipwrecks at depths that were previously undreamt of. The author takes the reader on the first-ever descent to the German battleship *Ostfriesland*, at 380 feet, which he made in 1990. He takes the reader on discovery dives to tankers, freighters, submarines, and World War One U-boats that lie off America's east coast. He evokes the thrill of what deep wreck-diving is all about - with insightful observations that can come only from one who has done it all.

Finally, Book Two covers in untold depth the successful 1994 mixed-gas *Lusitania* diving expedition - as well as the legal wrangles beyond. This behind-the-scenes look includes sinister elements that sound more like the plot devices of a grade B movie than abject reality: spies, sabotage, threats of lawsuit, even a bomb which destroyed a car in the author's driveway on the day before he left to testify in a federal hearing against the *Lusitania's* rival salvor. All the dirty details of the courtroom drama are included.

The Lusitania Controversies is a fresh, unexpurgated view of one of the most notorious shipwrecks in history. The *Lusitania* is a wreck whose infamy will not be forgotten.

THE LUSITANIA CONTROVERSIES
by Gary Gentile

BOOK ONE:
Atrocity of War
and a
Wreck-Diving History

5 - Whence

9 - Part 1 - Birth and Death of the *Lusitania*

40 - Part 2 - Fame Everlasting

70 - Part 3 - 1950's: The Origin of Wreck-Diving

89 - Part 4 - 1960's: Deep-Water Triumph and Turmoil

145-176 - Photographic Insert

208 - Part 5 - 1970's: Decompression Comes of Age

302 - Index   312 - Books by the Author

Gary Gentile Productions
P.O. Box 57137
Philadelphia, PA 19111
1998

Copyright 1998 by Gary Gentile

All rights reserved. Except for the use of brief quotations embodied in critical articles and reviews, this book may not be reproduced in part or in whole, in any manner (including mechanical, electronic, photographic, and photocopy means), transmitted in any form, or recorded by any data storage and/or retrieval device, without express written permission from the author. Address all queries to:

Gary Gentile Productions
P.O. Box 57137
Philadelphia, PA 19111

Additional copies of this book may be purchased from the same address by sending a check or money order in the amount of $25 U.S. for each copy (postage paid).

Website: http://www.pilot.infi.net/~boring/gentile.html

Picture Credits

All uncredited photographs were either taken or set up by the author. The front cover illustrations are from the author's collection: the top image is a contemporary postcard, the bottom image is a page from an unidentified contemporary magazine. Every attempt has been made to contact the photographers or artists whose work appears in this book, if known, and to ascertain their names if unknown; in some cases, copies of pictures have been in public circulation for so long that the name of the photographer or artist has been lost, or the present whereabouts are impossible to trace. Any information in this regard forwarded to the author will be appreciated. Apologies are made to those whose work must under such circumstances go unrecognized.

The author wishes to acknowledge Jon Hulburt, Drew Maser, and Gene and Joanie Peterson for reviewing the manuscript. Additional thanks are due to Eric Sauder for permission to reproduce photographs from his collection.

International Standard Book Number (ISBN) 1-883056-06-3

First Edition

Printed in Hong Kong

Whence

Underwater exploration has been one of mankind's most venturesome endeavors almost from the time when the species earned distinction from its pre-human ancestors. Finding the best places to fish must have occupied some portion of man's waking moments once he learned how much food was available in the clear fresh waters of inland lakes and rivers and, later, from the salty tumultuous surf. Sometime during his evolution he realized that he could *enter* the water in order to collect such comestibles as shellfish. Hand-dipping yielded to wading and eventually to breath-holding submersion, and skin diving was born - a description that was precisely accurate in prehistoric days when man wore nothing else.

Surely man must have wondered what secret caches lay hidden below the shimmering surface where sight did not readily penetrate. His inquisitive nature led him to explore ever deeper, and farther away from shore. Thus it can be said that part of man's heart has always been in the sea: that vast unknown realm where lurked the monsters of his imagination as well as the sustenance for his stomach and soul.

When we jump through the millennia to the time of recorded history, we find that man's enthusiasm for delving into the deep has not diminished. Quite the contrary, with the advent of modern technology and the achievement of greater feats, his enthusiasm grew and continues to grow. This growth has been fueled by two major objectives: military operations and the salvage of lost items and commodities. The cumbersome helmet rigs of yesteryear were adequate for the demands of such enterprise, but their utilization required a massive amount of surface support: air compressors, hoses, tenders, and large vessels from which to work. Only in the mid-1950's did self-contained underwater breathing apparatus become commercially available for recreational use.

It would not be incorrect to state that the wide-spread distribution of scuba has changed the world. With user-friendly equipment and professional training that is easily obtainable, nearly everyone can enjoy the sights and sounds of the lush liquid realm: from the cold dark waters of the temperate seas to the tropical coral reefs where fish gnashing their teeth is an ever-present chatter. Entire books have been written about the history of diving and the development of scuba; there is no need to repeat that saga here.

6 - The *Lusitania* Controversies

What follows is the chronicle of one particular aspect of recreational diving: the exploration of the plethora of sunken shipwrecks, leading up to the *Lusitania*. Nor does this pretend to be the complete story, for that cannot be written until every lost wreck has been discovered and surveyed. The forty years of effort recounted herein constitutes merely the infancy of the activity: the teething phase, if you will - and perhaps the onset of puberty.

Wreck-diving began almost as soon as tanks and regulators entered the public market. Scuba divers who ventured beyond the crashing surf zone soon found the remains of abandoned shipwrecks whose rotted timbers or twisted beams and hull plates provided artificial habitats for the denizens of the deep. But diving on these aged remains does not a wreck-diver make. Few hunters admire the beauty of the flowers in the field, or bother to take notice of the trees in the forest. Likewise, many divers fail to appreciate shipwrecks for their intrinsic merits. To them the wood or metal carcasses that litter the oceans' bottoms are man-made preserves in which game is concealed. Still today, the majority of divers who visit shipwrecks do so primarily to spear fish, catch lobster, collect edible mollusks, or swim around the site with little more than a passing interest.

For some divers, however, shipwrecks have a different and quite engaging appeal, perhaps a fascination, possibly even a zealous fervor. These people dive shipwrecks for the wrecks themselves and for what they have to offer in the way of personal challenge, exploration, history, and discovery. A true wreck-diver finds satisfaction in what he sees on a shipwreck, what he experiences during the dive, what he learns about the ship's past and its loss, and what he discovers - not only about the shipwreck but about himself and about the way he responds to diverse situations of uncertain outcome. Thus the exploration of shipwrecks and the challenges presented by the activity often lead to a healthy introspection and to an examination of self that far outweigh the more obvious motivations which outsiders are limited to noting. The excitement of a dive fills a gap in an otherwise dull and pedestrian life, and infuses the participant with confidence and the energy to face the tedium of the ordinary, workaday world.

By this description, wreck-diving is not a simple leisure-time activity - wreck-diving is an attitude. I want to stress the point that not everyone who dives on sunken wrecks is a wreck-diver, any more than a person who drives his car up the road to the summit of Mt. Washington is a mountain climber. Wreck-diving requires specialized skills and harbors a deep, abiding interest and fascination for overcoming obstacles in the exploration of long-lost ships.

The "wreck-divers" who meet these descriptive criteria are what this book is about: who they are, what they do, how they do it, why they bother.

When I set out to write the history of the *Lusitania* and a chronology of wreck-diving, I did not think that it would become so monumental a task. I wanted to tell it all: the good, the bad, and the ugly. Most of it is here in all its permutations. But the final story proved to be so long that it did not fit comfortably within the covers of a single volume that was feasible to print.

I had two choices: severely edit the material and leave out valuable detail and insights that would be lost to history forever, or tell it all and publish the work in two companion volumes. I chose the latter course.

I entered the wreck-diving arena in the wake of predecessors who had paved the way for the deeper and more stimulating quests to come. The contributions they made and the lessons they learned the hard way were passed on to subsequent generations of divers who benefited from the errors previously made and corrected, enabling them to make longer and faster strides in the development of the activity. This seamless transition is still going on. Like an endless treadmill it will never cease.

Because I have been so integral a part of the history and evolution of deep wreck-diving, it is impossible to tell that story without telling my own. In countless conversations and interviews I have been asked personal questions about my past: how I got into diving, do I have a death wish, have I ever been scared, and what led me to embrace wreck-diving with such dedicated enthusiasm? I have attempted to answer all those questions in the following pages.

In autobiographies that I have read, the authors wrote about their deeds and actions without relating their thoughts and feelings. Thus a valuable insight into their personalities was lacking, and a sense of sterility was pervasive throughout the book. Contrary to this custom, I have shown not only how I responded to situations, but why I responded the way I did and how I felt about events in which I was involved. When describing the behavior of others, I have related only observable conduct. This is not a novel, and I cannot presume to understand how others thought and felt, or what motivated their actions.

The chronicle that follows is a deep narrative history on many levels. I have chosen a style of retelling that juxtaposes the activity of wreck-diving with contemporary events and occurrences - personal and global. Otherwise the story would exist only in microcosm, as an isolated chronology without context. Instead I have woven a fabric with a complex pattern whose manifold overlays are stitched together by means of interconnecting threads that reach beyond the arena of wreck-diving.

This storytelling structure brings a broad picture of human history into focus. Wreck-diving does not exist in a vacuum, but in contin-

8 - The *Lusitania* Controversies

uum with contemporary events. In order to be fully understood and appreciated, it must be viewed against the backdrop of ongoing occurrences.

One condition of writing good fiction is creating strong characters. I have utilized my novelistic background and skills to accentuate those characters I have encountered in life who helped to make the world and wreck-diving what it was, and what it is today.

On the surface this book may appear to be too much my own story. In truth, however, it is everyone's story - seen through my eyes. Part of you is here. Part of everyone is here. The parts that you do not like you have the choice of changing.

Part 1
Birth and Death of the *Lusitania*

CONSTRUCTION AND DESIGN STATISTICS

Ships have lives not unlike people. They are borne by ways instead of wombs, they have skins of steel instead of flesh, they use an engine in place of a heart, and they circulate steam instead of blood. Once they are animated by the will of man, they assume an existence, a vitality, that is thenceforth self-sustaining. And just as people form unique identities as they age, so do ships forge their own personalities. The *Lusitania* captured the imagination of an era.

The *Lusitania* was conceived on the drawing boards of John Brown and Company, Clydebank, based on a design which was approved by the British Admiralty. The Cunard Steamship Company, owner and manager of the proposed ship and her running mate, the *Mauretania*, agreed to Admiralty specifications because of compensations offered by the British government and ratified by Parliament: low-interest construction loans and annual operational subsidies. This marriage between government and private enterprise proved advantageous to both parties. Cunard acquired a pair of modern liners at reduced cost and with less maintenance expense, giving the company a competitive edge in the transatlantic passenger service, while the Admiralty secured options on two high-speed vessels which in time of war could be quickly converted to armed troop transports or auxiliary cruisers and placed under naval control.

Admiralty specifications demanded that the two ships meet certain criteria: a minimum cruising speed of 24.5 knots, the emplacement of gun mounts and shrapnel shields, the installation of shell and powder magazines, and, for protection against enemy shell fire: engine rooms, boiler rooms, rudder, and steering gear situated well below the waterline, as well as high-rising coal bunkers along the boiler rooms (which comprised nearly half the length of the hull). The coal bunkers were so placed in order to provide protection against enemy shells. In addition, the majority of officers (excluding engineering officers) were required to be naval reservists.

Everything about the *Lusitania* was fabricated on a gargantuan scale. Upon launching she was touted as the "largest ship in the world." At 32,500 gross tons (38,000 displacement tons) she was the first vessel to exceed the tonnage of the *Great Eastern*, built by Isambard Kingdom Brunel nearly 50 years earlier, in 1858. (The *Great*

10 - The Lusitania Controversies

Eastern was scrapped in 1889.) The *"Lucy,"* as the *Lusitania* was affectionately called, boasted an overall length of 785 feet, a beam of 88 feet, a draft of 33.5 feet, and a height from keel to skylight of over 100 feet. From keel to funnel tops she rose 155 feet; her masthead towered 216 feet above her keel. Her anchors weighed 10 tons apiece; each link of her anchor chain weighed half a ton. She was steered by a rudder that weighed 56 tons; the stern post that supported the rudder topped 59 tons. The weight of her cast steel stem was over 8 tons.

Her massive hull was driven through the water by four bronze propellers, each turned by a steam turbine, the combined rating of which totaled 68,000 horsepower. At that time in the development of maritime machinery the standard means of propulsion was the triple expansion reciprocating steam engine. Turbines were strictly experimental; only half a dozen such power plants had been installed in other vessels, none of which were behemoths such as the *Lusitania* and *Mauretania*. Only after a good deal of testing on scale models in water tanks did the Admiralty and Cunard agree to go ahead with the implementation of this revolutionary technology. When in operation, 65,000 gallons of sea water per minute were needed to condense the steam after it passed through the turbines.

Steam to drive the turbines was generated by 25 boilers of which all but two were double ended (that is, furnace doors were fitted to both ends). The boilers were fired by coal at the prodigious rate of more than 40 tons per hour; each transatlantic passage consumed some 5,000 tons. Twenty-two trains were required to deliver the coal for a single passage.

The *Lusitania* did not have complete double hull construction throughout. Rather, the hull had a double bottom for practically the whole length, and for a considerable part of its length this double bottom extended up the sides to the waterline. Bulkheads divided her hull into 175 watertight compartments, making her "unsinkable" according to engineering conviction of the times.

INTERIOR DECOR

All passenger liners of the day were provided with libraries, writing rooms, smoking rooms (lounges), and large dining rooms. The *Lusitania* was no exception. Because British society was stratified by strict hierarchy, and different classes of people were not permitted to mix, first, second, and third class passengers read, wrote, smoked, and ate only with others of their own social status. The distinction between classes was not predicated upon race, religion, or sex, but by financial standing. A pauper could travel first class if he was willing to spend his life savings for the privilege. Otherwise, it should be noted that a great disparity existed between the quarters for steerage passengers and the description of a liner's most salient features as advertised.

Of greater significance to her rich contemporary patrons was the elegance of accommodations. She could carry 552 first-class passengers in extravagant grandeur in compartmented staterooms, 460 second-class passengers in less opulent cabins, and 1,186 third-class passengers in Spartan settings. The latter were better known as "steerage" passengers because they generally occupied crowded bunk rooms aft, although the *Lusitania* provided space for many of them forward on the lower decks. In a manner of speaking one could say "the higher the deck the higher the society."

The *Lusitania* was appointed sumptuously and finished in rich woods and ornate decorations. The dining saloon was a two-deck affair boasting a circular balcony and topped by a domed plaster ceiling of intricate design; 500 people could be seated in Edwardian era splendor. The public rooms were commodious, each paneled with its own motif of carved moldings and pilasters, and adorned with embroidered valances, silk curtains, leather upholstered chairs, mahogany furniture, inlaid tables, and deep pile carpeting or shining linoleum. High ceilings or arched domes of leaded glass added an ambiance of open-air spaciousness. For the wealthy, the *Lusitania* was a floating palace complete with exquisite food, rooms, and service. Steerage passengers dined on simpler fare in cafeteria style, but that was better than the old days when they had to supply their own victuals and eating ware.

In keeping with modern technology, the *Lusitania* was equipped with a long distance communication device known as a Marconi wireless. Short for "wireless telegraphy," this recent invention enabled ships at sea to communicate by Morse code with land-based stations and with other ships that were likewise equipped. Range was limited in early models to a few hundred miles, so transatlantic vessels could be out of communication for days at a time unless their messages were intercepted by another ship and retransmitted in leapfrog fashion.

LAUNCHING AND MAIDEN VOYAGE

Twenty thousand people witnessed the *Lusitania's* launching, on July 7, 1906. After a year in the fitting-out basin followed by sea trials to test her machinery, she began her maiden voyage on September 7, 1907; more than one hundred thousand singing, cheering people thronged the docks and riversides to watch her first departure. Her accommodations were filled to capacity with more than 2,000 passengers, to the chagrin of another two hundred people who expected to board at Queenstown, her first stop.

The British harbored great expectations that, in addition to being the biggest and most sumptuous liner in the world, she would be the fastest. They must have been horribly disappointed when she failed to break the transatlantic speed record, leaving the *Kaiser Wilhelm II* of Norddeutscher Lloyd still the champion. The *Lusitania* completed the uneventful crossing in five days fifty-four minutes, at an average speed of 23 knots. Not only was this less than her design speed but less than the Admiralty's requisite speed. However, due to fog and stormy weather encountered along the way, Captain James Watt wisely reduced revolutions to make a slower but safer passage. The ship actually attained speeds in excess of twenty-five knots when conditions permitted.

The great liner was first to enter New York harbor via the newly completed Ambrose Channel, which the Army Corps of Engineers had just finished dredging. Scores of vessels of all descriptions blasted their whistles as the *Lusitania* hove into view. Throngs of people numbering tens of thousands greeted her with enthusiasm; they stood in awe of her monstrous proportions. So enthralled by her dimensions was one reporter that he described her as "a skyscraper adrift." For days, thousands of sightseers purchased boarding tickets so they could walk the *Lusitania's* holystoned decks. For years they would boast about the experience.

Race for the Blue Ribband

The *Lusitania* soon redeemed herself and restored the faith held in her by loyal British subjects. On her second westward passage she surpassed the highest daily speed ever attained by the *Kaiser Wilhelm II*, and shortened the overall time of crossing to four days nineteen hours fifty-two minutes, thus capturing for herself and for the honor of Great Britain the coveted Blue Ribband, a symbolic trophy which represented the attainment of the fastest Atlantic crossing. The experimental steam turbine had proven its worth. Shortly thereafter she claimed the fastest eastward passage as well.

Sister Ship *Mauretania*

If the *Lusitania* lived up to her promise, the *Mauretania* surpassed hers. Sister ships are never identical twins, even when they are built at the same shipyard; there are bound to be differences in design and decoration. The *Mauretania* was built by Swan Hunter and Wigham Richardson, Newcastle-upon-Tyne. Her interior ornamentation (or ostentation, depending upon one's perspective) found differing favor among frequent travelers, some preferring one ship to the other. Steerage passengers seldom cared; their primary interest was in getting somewhere, not in how they traveled. Most of them were emigrants.

Because construction of the *Mauretania* was begun after that of the *Lusitania*, minor improvements were made to the *"Mary's"* machinery and boilers. She was fitted with four-bladed propellers instead of three-bladed, and the blades were half a foot longer. All this culminated in greater efficiency which translated to a slight superiority in speed. Furthermore, the Scottish built ship was a wee bit larger. On her maiden voyage she set a new speed record for the amount of miles steamed in a single 24-hour period, but, as with the *Lucy*, fog and foul weather prevented a quick coup for the Blue Ribband. However, despite clinging fog and gale force winds encountered during her return, the *Mary* managed to best the *Lucy's* eastbound time by twenty-four minutes, thus wresting the trophy from her slightly older twin. Fame is often fleeting.

The *Lusitania* managed to take back the Blue Ribband from the *Mauretania*, but only for a short-lived respite. For a couple of years the two ships played a game of besting each other. Then the *Mauretania* seized the trophy once and for all and held onto it for the next twenty years. Both ships eventually exceeded their design speeds, usually averaging better than 25 knots on crossings, sometimes sustaining 25.5 knots. Once the *Mauretania* made over 26 knots all the way across. The *Lusitania's* fastest passage was 25.88 knots.

Early Career

Although both liners were loved by the company and the public alike, they had their share of problems. One which was shared by both was excessive vibration at high speed. The aft end shook so much that reading and writing were difficult and, for those less fortified, sickening. Massive alterations were required. Much of the *Lusitania's* after interior was redesigned to accommodate the necessary structural modifications: additional beams and strengthening members that had to be hidden from the eyes of her passengers.

The two ships provided such reliable service that together they came to be known as the "Atlantic ferry." Weather occasionally moderated their pace, as did unforeseen mechanical difficulties, both to be expected and both shared by all vessels afloat. Each experienced significant events.

During her first year of service the *Lusitania* carried two shipments of gold from England to the U.S.: one worth $10 million, the other $12 million.

Almost as a prelude to her final demise, in 1908 the *Lusitania* was held up by shellfire. Fort Hamilton and Fort Wadsworth, which guard the entrance to New York harbor, were engaged in gunnery practice when the liner left the Cunard pier and slipped down the river toward the lower bay on her way to Liverpool. As she passed through the Verazanno Narrows, dummy shells from both forts crisscrossed in

14 – The *Lusitania* Controversies

front of her and raised huge splashes in the water. Each 12-inch shell weighed more than half a ton. The concussion from the discharging guns wreaked havoc in nearby communities, smashing crockery and blowing out window panes. The *Lusitania* escaped unscathed after a delay of half an hour.

Also in 1908, in an effort to increase speed and efficiency of operation, the *Lusitania's* three-bladed propellers were replaced with four-bladed props like those that proved so effective and economical on the *Mauretania*. It was this change that enabled her to beat her sister's record temporarily. The following year, two blades of one of these new propellers snapped off in mid ocean. The ship shook violently for several minutes until engineers diagnosed the problem and shut down the engine turning the damaged propeller. She completed her passage to New York at reduced speed, then upon her return to Liverpool she was taken out of service and dry-docked for repairs.

Stormy Encounters

Even ships as large as the *Lusitania* suffered from foul winter gales. In November 1909 she buried her bow to such an extent that cargo boom cradles were twisted off and three glass windows in the bridge superstructure were "splintered into a thousand bits." In December 1909 a wall of green water struck the bridge superstructure with such force that it carried away the chartroom ladders, smashed the bridge railing, and stove in the shutters on the chartroom windows. For a moment, "the officers on the bridge were up to their boot tops in water. A second wave smashed through the skylight over a stateroom.

In January 1910, the *Lusitania* was nearly swamped by a wave estimated at *eighty* feet in height. Not only were the bow and entire forward deck engulfed, but "the wall of water went higher than the wheelhouse," then "swept down her decks the whole length of the ship." The impact stove in the superstructure's steel plates. Water smashed through the shutters, unshipped the wheel, and carried the wheel and the helmsman "across the room and up against the chartroom partition." Chief Officer McNeil was standing on the bridgewing when he saw the wave approaching; he dashed for the wheelhouse but hadn't quite made it when the monster wave struck and washed him in through the doorway. Bridge boy Tommy Hughes flattened himself on the deck of the exposed bridge, was washed away by the force of the water, but managed to grab hold of an iron stanchion as he was being swept overboard. On top of the wheelhouse, Third Officer Storey ducked and clung to the compass stand until the engulfing sea passed.

Farther aft, ladders were ripped loose from their mounts, doors were stove in, glass was shattered, davits were twisted and bent, two starboard lifeboats were lifted from their chocks and demolished against bulkheads, and the officers' quarters were flooded. Astonishingly, although minor injuries abounded, no one was killed.

Electrical shorts plunged the wheelhouse and chartroom into darkness. The navigational lights were extinguished. Some of the contents of the chartroom were washed out, including the log. The bloodied bridge watch struggled in darkness and waist-deep water to restore order. For forty minutes the liner was left to the mercy of the seas, until she got under way at a speed of ten knots. Control was managed from the auxiliary steering station until the bridge wheel could be reshipped. She was a sad looking ship when she arrived in New York late, but better than never.

SUCCESSFUL OPERATION, MECHANICAL TRIBULATIONS, AND OPPORTUNE RESCUE

The *Lusitania's* most successful year of operation was 1911, during which time she transported over 40,000 passengers and delivered more than 78,000 bags of mail. In response to the sinking of the *Titanic* (on April 15, 1912) four lifeboats were added to the *Lusitania's* usual complement of sixteen. The extra capacity left her only four hundred seats short in the event of an emergency.

In June 1912 came the first of a series of engine breakdowns that plagued the *Lusitania* for more than a year. First, the port low-pressure turbine became disabled, and for four months the ship operated with the propeller on that shaft feathered to reduce drag. She was laid up for repairs for a couple of weeks in November, then returned to service.

Barely a month passed before a freak accident occurred. A small piece of marline fell into the telemotor and jammed the steering mechanism just as the helmsman ported the helm to avoid an oncoming steamship. With collision imminent, "the turbines were sent full speed astern without the usual precaution of stopping them first and then reversing them slowly." The collision was avoided, but the port turbine was damaged in the procedure and would no longer move. The newly installed blades were twisted out of shape, and when further inspection was made, the blades on both starboard turbines were found to be bent just as severely. This necessitated a complete overhaul of the three turbines, and the replacement of 1,250,000 blades. The parts and labor costs were half a million dollars, not counting lost revenues for the eight months that the ship was out of service. The port high-pressure turbine was not rebuilt. Once again the propellers were changed, this time so the blades rotated outboard instead of inboard.

Her reincarnation debut in New York did not occur until August 29, 1913. The speed of her first crossing after being re-engined was

16 – The *Lusitania* Controversies

disappointing; Cunard explained that of the two hundred stokers, seventy-five were greenhorns who were not well acquainted with their tasks. With each succeeding passage her speed increased until she reduced her time to four days eighteen hours and five minutes: not a record, but a respectable time nonetheless.

Of more significance than the speed of passage was the speed with which members of the *Lusitania's* crew volunteered to man a lifeboat in order to rescue the crew of the foundering schooner *Mayflower*. For their bravery, the rescuers received gifts and monetary rewards in addition to testimonials.

The Wages of War

Then came 1914 and the beginning of the war with Germany. The Admiralty exercised its option with Cunard by taking over the *Mauretania*, which served throughout the war first as a troop transport, then as a hospital ship, and again as a transport. But the *Lusitania* was permitted to operate commercially.

During her first return passage after the declaration of war, the *Lusitania* encountered what was believed to be a German naval vessel in the North Atlantic. Captain David Dow ordered immediate evasive maneuvers that took the *Lusitania* south into a fog bank. She gradually swung east and north around the presumed enemy vessel, and docked in Liverpool late but unharmed.

Ticket sales then dropped dramatically. Scheduled round-trip crossings were reduced to one per month for the *Lusitania* (compared to sixteen voyages in 1911), and this at a time when the *Mauretania* was chartered for war work and was no longer engaged in transatlantic passenger service. Even so, paid fares did not generate enough income to cover the expense of operation. In order to cut costs, six of the *Lusitania's* twenty-five boilers were shut down; this saved money on expenditures for both coal and labor. Greater economy plus the Admiralty subsidy (Great Britain still needed mail service to and from the States) enabled Cunard to barely break even.

With 25% less steam, the *Lusitania* was able to make at most 21 knots, yet she was still the fastest liner on the North Atlantic run. (Her German competitors were either interned in neutral America, confined to German ports, or roaming the high seas as merchant raiders.) This speed was deemed sufficient for her to proceed unescorted since the Imperial High Seas Fleet was bottled up in the North Sea by Britain's Grand Fleet, and since she could easily outrun a lurking enemy submarine (hereafter called a U-boat, in accordance with Winston Churchill's definition, to distinguish it from an Allied undersea craft). Despite the provision made for emplacing guns, the *Lusitania* was never armed. Her sole means of defense from enemy attack was her speed. Protection for her passengers and crew was guaranteed by provisions of the Hague Convention of 1907 regarding the legitimate destruction of merchant ships and safe conduct for non-combatants.

The Hague Convention and International Law

International law provides for a humane approach toward war, an oxymoron if there ever was one. Briefly, "Vessels belonging to the enemy state, and notably warships, may be attacked, captured, or destroyed by a belligerent man-of-war anywhere on the high seas or in the territorial waters of the contending belligerents, at any time and without notice. But enemy merchantmen are not to be subjected to such summary and drastic treatment." Vessels "engaged in peaceful commerce and other pacific activities" are differentiated from warships because "an enemy merchant ship may actually belong to a class of vessels exempted from capture and destruction by special conventions and usage," such as hospital ships, and "enemy merchantmen may have neutral persons and neutral cargoes on board." A neutral cargo is one that has no military value; a vessel carrying troops or war materiel does not qualify as neutral if she is trading with a belligerent country.

Difficulties arose with the substantiation of neutrality. An onboard inspection was required in order to ascertain the nationalities of the passengers and crew and to identify the nature of a vessel's cargo. In pre-wireless and pre-U-boat days, the accepted procedure required that a warship signal an unidentified merchantman to heave to: first by flag or blinker light, then by a shot over the bow if she failed to comply. The warship sent a boarding party to the stopped merchant ship to examine her clearance papers and manifest. If suspicion was aroused by seemingly doctored documents, an actual scrutiny of the holds was conducted, but this was a rare event.

Neutral ships traveling between neutral ports were released with apologies. Neutral ships bound to or from a belligerent country were released only if their cargoes contained no contraband. If the ship was registered to a belligerent country or if contraband was found aboard, she could then be sunk or seized as a prize of war. In *any* case, the safety of the passengers and crew was ensured by the Hague Convention.

Ideally, passengers and crew were taken aboard the warship; citizens of neutral nations were surrendered at a neutral port according to military exigency, while citizens of belligerent nations could be held as prisoners of war. Under circumstances less than ideal, people were given time to abandon ship in lifeboats, the boats were adequately stocked with water and provisions, and medical attention was provided for the injured. Even so, a lifeboat left adrift in mid-Atlantic was no bargain. Many sailors were never seen again despite receiving this "humane" or "gentlemanly" treatment.

The latter strait was the best that a U-boat could offer. It had neither the manpower to designate a prize crew nor space aboard to accommodate large numbers of survivors.

Disputes sometimes arose between a merchant ship captain and the boarding officer over what commodities constituted justifiable contraband. Guns and ammunition obviously fitted the category. But Germany expanded its classification system to include leather that could be used for making saddles for the cavalry, copper that could be alloyed with other metals to make brass shell casings, and food that might be supplied to troops fighting German soldiers on the western front - in fact, nearly every conceivable product could, by this thwarted Teutonic reasoning, be construed as contraband. This German policy effectively prevented all free trade between neutral countries and Allied nations. Disagreements at sea were arbitrated by the vessel with the greatest amount of fire power.

The darker side of mediation at sea was the scuttlebutt about German cruelties. Stories abounded in which merchant ship survivors were gunned down in their lifeboats or while clinging precariously to life preservers in the frigid water. The crew of the *Mimosa* was reportedly lined up on the deck of the attacking U-boat, which then submerged, leaving the men to drown.

Merchant Vessel Protective Measures

Since posturing to U-boats did not always work the way it was supposed to, some merchant marine captains took it upon themselves to save their ships from possible destruction by steering away and making off at high rpm's - a reasonable response under the consequences that often eventuated from meek submission. Some ships fired distress rockets or, if equipped with wireless, signaled for help. All attempts at self preservation were considered by Germany as acts of aggression, thereby entitling U-boats to fire upon offending merchantmen and escaping passengers and crew.

Great Britain retaliated against what it perceived to be German violations of the Convention by issuing official orders to merchant marine captains to flee or ram, as deemed appropriate by the situation, and to haul down the British flag and fly the flag of a neutral nation in order to avoid close scrutiny. In February 1915, the *Lusitania* was returning home when Captain Dow received a wireless communication from the Admiralty that U-boats were lurking in the Irish Sea. He promptly raised the American flag and proceeded at full speed to Liverpool. The flag might make a U-boat commander hesitate just long enough for his indecision to prove his undoing, and allow the liner to get away. After this incident, Captain Dow was given shore leave in order to recuperate from the stress of command. He was replaced by Captain William Turner.

Britain armed all its tankers and freighters, usually with a deck gun positioned aft so it could fire at an attacking warship while the merchantman attempted to flee, and sometimes with machine guns and other small arms. Naval gun crews were provided free of charge. Since armed merchantmen were classified as naval auxiliaries, according to the Convention they were subject to attack without warning.

British Offensive Measures

Compared to traditional warships, U-boats were disadvantaged by their slow speed, thin plating, and limited armament. A single shot through the pressure hull could prevent a U-boat from submerging, its primary means of defense and escape. The only superiority U-boats had over surface craft was stealth: they could slip through the sea undetected by prowling enemy patrol craft. A U-boat in full view on the surface was no match for any kind of warship as it was easily outgunned and outmaneuvered.

The British equipped harmless looking merchantmen with guns concealed behind false bulwarks which could be dropped at a moment's notice to permit a clear field of fire on unsuspecting U-boats as they approached to examine her papers. These Q-ships, as the decoy vessels were called, created an atmosphere of suspicion among U-boat commanders, compelling them to launch torpedoes from a respectable distance rather than to take the chance of falling into a trap. Because of a U-boat's weaknesses and by the nature of its method of attack, it could not wage war effectively according to Convention.

Escalation

Britain stemmed the flow of German seaborne commerce by patrolling the North Sea, the entrance to the Skagerrak, and the English Channel, thus blockading Germany most effectively. German merchantmen had hardly any chance of evading the thick screen of destroyers, drifters, and motor patrol boats.

Germany declared a war zone around the British Isles. This meant that vessels of any nation, whether neutral or belligerent, no matter what their cargoes, were liable to attack without warning by German warships, U-boats, and airplanes, if those vessels were encountered in or near British home waters. On the high seas, a U-boat might abide by the rules of the Hague Convention if such action posed no threat of trickery or counterattack.

During nighttime forays, Britain sent minelayers into German waters to sow secret devastation off German and German-held ports. Britain also initiated a large-scale operation to stretch a barrier of mines across the North Sea between the Orkney Islands and Norway. This restricted traffic to specified lanes of travel, knowledge of whose zigzag courses were entrusted only to vessels of Allied nations and to vessels of neutral nations that did not trade with Germany. Submerged submarine nets spanned the gaps between concentrations of mines. This barrage of explosives and nets denied to Germany safe passage of the North Sea. Taken in conjunction with the danger of negotiating the English Channel, which was heavily patrolled, German ports were essentially isolated from the rest of the world, forcing a massive overland importation of goods.

Germany countered this initiative with site-specific mining. Surface minelayers by night and submarine minelayers by day dropped clusters of submerged mines - some with time-delay mechanisms that prevented them from arming for several days or weeks - immediately outside British ports or in the lanes leading to those ports. British mine sweepers worked overtime to keep the port approaches cleared for traffic, while armed merchant trawlers deployed submarine nets and dropped depth charges on U-boats that got caught in the nets, then fired shells at them when they were blasted to the surface.

U-BOAT - THE DREADED ADVERSARY

Germany's final countermeasure against the various Allied stratagems was the introduction of unrestricted submarine warfare: submerged attack against any ship, anywhere, at any time. Neutral nations objected to unrestricted submarine warfare since their ships were being sunk in a war in which they were not politically involved. Also a hazard to neutral shipping were the mine fields laid by Germany off the British east coast, and the British blockade of German ports and the mining of the North Sea. Through the complex organization of maritime commerce the entire world was embroiled tangentially in the European conflict by the draconian conduct of the active belligerents.

The British attitude toward submarines had always been one of derision: they were perceived as defensive weapons only, to protect a country's coast, and any offensive use was considered invectively as "underhand, unfair, and damned un-English" (in the words of the First Sea Lord, Admiral Fisher). Be that as it may, fair play has no place in war, as Fisher himself once pointed out: "The essence of war is violence; moderation in war is imbecility."

The Birth and Death of the *Lusitania* – 21

Germany planned its U-boat offensive as early as 1908, when Vizeadmiral Freiherr von Schleinitz wrote, "To sink a large number of English merchant ships would be much more significant than defeating an opponent in a naval battle." Not everyone in the German navy agreed with him, and few recognized the vast offensive potential of the undersea threat. But despite intradepartmental opposition, Germany initiated a major research, development, and construction program in preparation for the day of expansionism that the Kaiser knew was coming.

So while the British were smug in their denunciation of submarines as worthless escorts to the navy's fighting force, and while the German surface fleet was kept completely at bay, U-boats roamed the seas with virtual impunity and waged a vigorous offensive campaign that proved remarkably - or disastrously - successful.

Tonnage statistics were little less than astounding. In 1915, U-boats sank more than one million tons of Allied merchant shipping. This translates to an average of between fifty and one hundred vessels *per month*. To insure the *Lusitania* for $10 million against loss due to marine peril or casualty of war, Cunard had to pay her underwriters $50,000 per voyage.

This is the point to which the Great War had escalated by the time the *Lusitania* was prepared to depart on what proved to be her final voyage, on May 1, 1915.

DIRE PREDICTION

Fiction writers are wont to infuse feelings of impending doom into the dialogue of their characters, a plot device that has no place in an historical account. In the *Lusitania* circumstance, the passengers experienced more than a *feeling* of impending doom; there was actual documentation of it. On the same day that the *Lucy* departed New York, the morning editions carried a warning from the German Government. Not only did this warning appear on the same page as Cunard's Atlantic service advertisement, in which the *Lusitania's* departure date and time were capitalized and boldfaced so they stood out from the scheduled departure dates and times of the other Cunard liners, but in some newspapers the warning was published adjacent to the ad. It read:

NOTICE!

Travellers intending to embark on the Atlantic voyage are reminded that a state of war exists between Germany and her allies and Great Britain and her allies; that the zone of war includes the waters adjacent to the British Isles; that in accordance

with formal notice given by the Imperial German Government, vessels flying the flag of Great Britain, or of any of her allies, are liable to destruction in those waters and that travellers sailing in the war zone on ships of Great Britain do so at their own risk.
IMPERIAL GERMAN EMBASSY,
Washington, DC, April 22, 1915

The warning does not appear to have frightened anyone into changing reservations. The few last minute cancellations were routine: the result of passengers making different arrangements for personal reasons having nothing to do with the notice, and generally due to their own scheduling priorities. There was discussion of the matter, to be sure, as several survivors later recalled. But the talk was not heated and was made more out of curiosity - or indignation - than out of fear.

Tugs pushed the great ocean monarch out of her dock and into the steady flow of the Hudson River. As her boilers poured out steam for the turbines the *Lusitania* gradually picked up speed, until she was moving completely on her own and slowly outdistancing the tugs. Nothing appeared out of the ordinary. She soon reached the open ocean beyond Sandy Hook, dropped off the pilot, then proceeded into the broad Atlantic and shaped a course for an unintended destination and a direful date with destiny.

The *U-20* and Lieutenant Schwieger

Meanwhile, on the other side of the Atlantic, the German U-boat *U-20*, under the command of Kapitanleutnant Walther Schwieger, circled Ireland counterclockwise from the north and west, patrolled near land along the southern shore, then steered northeast into the Irish Sea toward St. George's Channel on the way to Liverpool, preceding the *Lusitania's* anticipated course. After several unsuccessful attacks, on May 5 Schwieger stopped the small schooner *Earl of Lathom*, ordered her crew to abandon ship, and sank the vessel by shell fire. The five survivors rowed to shore at Queenstown and reported the incident to authorities.

Later that day, the *U-20* launched a torpedo at the British steamer *Cayo Romano*. The torpedo missed and the ship got away. The *Cayo Romano* arrived safely at Queenstown and filed a report that reached the Admiralty a few hours after that of the *Earl of Lathom*. Both attacks took place within sight of Old Head of Kinsale on the southern point of Ireland.

During the night and early morning the *U-20* patrolled slowly northeast along the *Lusitania's* projected course. In the vicinity of the

Coningbeg lightship, which marked the entrance to St. George's Channel, the fog lay patchy and sometimes thick. Despite poor visibility, Schwieger encountered the British steamer *Candidate* and convinced her crew by shell fire to abandon ship, after which he launched a torpedo at the motionless target and struck her squarely. The *Candidate* refused to sink in a timely fashion so, economizing on torpedoes, Schwieger finished her off with shell fire.

Shortly thereafter came the British steamer *Centurion*. Schwieger launched a torpedo without warning. The merchantman was struck and brought to a halt, but proved to be as reluctant to sink as the *Candidate*. After the crew abandoned ship, Schwieger moved in for the coup de grace, this time delivering the fatal blow with another torpedo.

There were no fatalities on either the *Candidate* or the *Centurion*. All survivors were picked up by patrol craft later that day. Again reports were sent to the Admiralty about U-boat activity in the area. In response, the Admiralty issued warnings by wireless to ships in the vicinity and to those arriving and departing that U-boats were operating in the Irish Sea. These warnings were received on the *Lusitania*, with the additional admonition to avoid headlands, to pass harbors at full speed, and to steer a mid-channel course.

The *U-20* reversed direction and proceeded southwest along the Irish coast. Although Schwieger made no mention in his war diary, which was written after the cruise and after the heat of controversy was ignited, it has been suggested that he was operating under orders to linger in the area to await the arrival of the *Lusitania,* whose course was a long established routine and whose times of arrival and departure were advertised in the newspapers for everyone to read. Transatlantic liners had been following the same route for decades, except on rare occasions when they rounded Ireland to the north. However, if the *Lusitania* had been targeted as Schwieger's primary mission, it is unlikely that he would have announced his presence so flagrantly by attacking vessels of lesser importance, thus relinquishing the advantage of surprise.

FATEFUL ENCOUNTER

On May 7 the *Lusitania* approached the south coast of Ireland enshrouded in fog. Captain Turner prudently reduced speed to 15 knots until the cloying mist evaporated and clear skies reigned overhead, at which point he increased speed to 18 knots. At this time the *Lusitania's* lifeboats were swung out, watertight bulkheads through which passage was non-essential were closed, and portholes were dogged down tight.

Wireless messages received from the Admiralty warned that U-boats were operating off the south coast of Ireland and that one had been spotted some twenty miles south of the Coningbeg lightship.

Captain Turner took a sighting on land to establish his position; it was easy for a vessel to drift off course during a couple thousand miles of ocean travel. He identified what he thought was Brow Head. He turned slightly to port, spotted what he thought was Old Head of Kinsale, turned slightly to starboard, then steamed along the coast at a distance of ten to twelve miles while his officers took a four-point bearing to establish without doubt the *Lusitania's* position.

On the *U-20*, Schweiger observed a vessel approaching on a nearly reciprocal course. Evidence that his war diary was altered to save face in light of subsequent disputes is implied by these entries: "Sight dead ahead four funnels and two masts of a steamer steering straight for us." He could not possibly count the number of funnels and masts from straight ahead. He would see only the ship's bow with the masts and funnels in line and appearing as one.

This statement further conflicts with the next entry which describes diving to periscope depth and proceeding "at high speed on an intercepting course" and "steamer alters to starboard . . . so permitting an approach for a shot." Needless to say, a steamer "dead ahead" and "steering straight for us" is already on an intersecting course.

Log entry times state that fifty minutes passed between the sighting of the liner and the launching of the torpedo. During that time the *Lusitania*, proceeding at 18 knots, would have traveled 15 nautical miles. This implies that when Schwieger first spotted the *Lusitania's* funnels she was more than 17 statute miles away.

These apparent contradictions neither affect nor justify the outcome, but perhaps they indicate the change in Germany's political posture from an attitude of triumph and arrogance to one of tragedy and conciliation. The war diary could have been altered in order to placate a disturbed American public.

The question begged by the questionable enlightenment noted above is whether the *Lusitania* unwittingly contributed to her own demise by turning away from land and into the *U-20's* field of fire.

Philosophical musings notwithstanding, according to Schwieger's war diary and wireless transmission records, he launched a single torpedo from a distance of seven hundred meters (nearly half a mile). Lookouts on the *Lusitania* spotted the track too late for the ship to avoid the fatal impact. The torpedo struck the liner on the starboard side slightly abaft the bridge or between the first and second funnels. A dull thud was accompanied by a large spout of water. A moment later, another and larger explosion occurred farther aft between the second and third funnels.

The Controversial Second Explosion

Some witnesses aboard the *Lusitania* claimed to have seen the track of a second torpedo. Schwieger observed the secondary explosion, and later wrote tersely in that regard: "boilers, coal, or powder?"

Hot boilers submerged in cold water commonly exploded when a ship sank while steam was still under pressure. Thus it was general practice to blow off steam before abandoning ship. In this instance, the second explosion occurred almost instantly, long before the sea could have flooded the boiler rooms extensively enough to cause such a reaction.

Coal dust explosions in steamships were also well-recognized occurrences. They are similar in nature to straw dust explosions in barns and warehouses and grain dust explosions in silos. As the *Lusitania* neared the end of her passage, her supply of coal was nearly exhausted and her bunkers were largely empty, leaving behind a thick residue of dust. The torpedo's detonation shook the ship violently, flinging the fine black motes from the deck into the air. The flash and flame of expanding gases and sparks from metal striking metal could easily have ignited such airborne dust with the devastating result observed.

Despite contemporary and continued allegations, the Admiralty did not then conceal nor later try to cover up the fact that on her final passage the *Lusitania* carried ammunition in her holds. The military portion of the cargo consisted of 4,200,000 Remington .303 rifle cartridges, 1,250 cases of unfilled shrapnel shells (casings only), and 18 cases of fuses: 173 tons of munitions in all.

Germans and conspiracy advocates would have it that the shrapnel shells were filled with fulminate of mercury - a highly volatile explosive - but filled or not, none of the ammunition was stowed anywhere near the site of either explosion. The forward magazine was located forward of the bridge (nearly 100 feet from where the torpedo struck), the after magazine was located abaft the engine (nearly 300 feet from the site of the second explosion). Coal bunkers were the only below-waterline compartments situated along the length of the hull where the two explosions occurred. For obvious reasons, munitions were never stowed in coal bunkers or on the open decks above them, where a spark or a carelessly tossed cigarette could have ignited the whole batch.

Not to be dissuaded by logic and common sense, conspiracy advocates claim that a liner like the *Lusitania*, constructed with a double hull and watertight compartments, could not have sunk as fast as she did without the aid of additional explosives - perhaps secreted in wooden boxes disguised as crates of fruit, and listed on the manifest as such. They ignore similar instances: the *Empress of Ireland* settled

26 - The *Lusitania* Controversies

to the bottom only fourteen minutes after collision; the *Britannic* (sister ship of the *Titanic*) disappeared completely within twelve minutes of striking a German mine; the *Arabic* went down nine minutes after being torpedoed. No charges of hidden explosives were made against those vessels.

Furthermore, the magnitude of the second explosion was nowhere near large enough to have resulted from the detonation of 173 tons of munitions - or an appreciable percentage thereof - and was especially too small to have been the instantaneous discharge of a moderate quantity of high explosive such as fulminate of mercury. Remember the *Maine* - the U.S. battleship that blew up and sank in Havana Harbor in 1898 when her magazine exploded. The explosion demolished nearly half the vessel, and the ship sank in a matter of minutes.

In the case of the *Lusitania*, most witnesses characterized the second blast as similar to the first, only larger. In addition, the explosion was not described as one of internal origin, but, by the nature of spume and spray that engulfed the superstructure, one that was manifested against the outer hull.

Schwieger must have demonstrated considerable restraint if he did not, as he insisted, launch all his remaining torpedoes at the largest enemy vessel he was ever likely to encounter, knowing as he must that there was very little chance that a single warhead could have any effect on an enormous four-stacked vessel other than to slow her down, anger the passengers and crew, affront the Admiralty, and earn jeers of derision from U-boat command for letting such a magnificent target escape. He alleged to have been saving his last two torpedoes for targets he might engage on the way back to base: an economy that would have been a poor trade for a coup like the *Lusitania*.

His war diary also states, somewhat incongruously, that after several minutes of observation through the periscope, "I could not have fired a second torpedo into the crowd of people struggling to save their lives." That sentiment had not prevented him from firing the first torpedo.

Schwieger's war diary goes on to state that positive identification was made before leaving the site by reading the name off the bow through the lens of his periscope. Yet Turner testified that the name and port of registry were painted out.

SINKING AND PAINFUL DEATH

The *Lusitania* took a list to starboard and inclined down by the head. Captain Turner immediately issued emergency orders: to close the watertight doors that remained open, to steer the ship toward land, to heave to by stopping and reversing the engines, and to prepare the lifeboats for launching but to wait until sufficient way was off

The Birth and Death of the *Lusitania* - 27

before lowering the boats. The liner began a long circular sweep to port, but whether the order to decelerate was acknowledged by the engine room was not ascertained in court. The men in the boiler rooms were soon chased from their stations by the quickly rising flood; some did not escape and were drowned. The absence of stokers and fires doused by sea water meant that no steam was available for the engines.

The fast increasing list made other tasks undoable. Steam lines burst, generators failed, motors stopped, power leads short-circuited. The mechanical life that was the *Lusitania's* soul began disintegrating in an ever-quickening rush.

Passengers swarmed to the upper decks in evident distress but generally without panic. They rushed to the lifeboat stations where crew members stood by awaiting the order to lower away. Because of the reduced number of passengers and crew aboard, more than enough space existed to accommodate them all in the boats. But due to the immediate pronounced list, the port boats could not be lowered. Their weight pressed them against the sloping hull and caused them to scrape haltingly over the riveted steel, which quickly stove in their sides.

Abandon ship procedures did not go smoothly. Turner was marking time for the ship to slow to a safe speed for launching, so he never gave the order. Officers and crew soon took it upon themselves to lower away in a necessarily undisciplined manner - with disastrous results. The *Lusitania's* high speed caused the first boat to capsize as soon as it touched the water; all the occupants were dumped overboard into the frigid unbreathable brine. Another boat was upended when a seaman let go the falls while the other rope was still secured; everyone was pitched into the sea from the height of the deck, to crash atop each other.

Some of the boats that survived launching had no plugs for the drains; they flooded quickly and swamped. Some had no bailing buckets. Some had no oars. The canvas sides of the collapsible boats could not be properly raised. Nor was there time to correct the problems as there should have been. The *Lusitania* was settling fast.

Events telescoped with alarming speed. Less than half the lifeboats got away safely, and most of those were not filled to capacity. People who waited too long to abandon ship soon found the deck disappearing beneath them.

The bow crashed into the bottom some 300 feet beneath the surface of the sea. Hundreds of people were swept off the ship as human flotsam, most to die the numbing death of hypothermia. So long was the hull that the stern still rose above the surface, with survivors clinging to the after rails and superstructure. Captain Turner was swept off the bridge as it plunged beneath the waves.

If these events seem too fast in the telling, consider the time frame. Eighteen minutes after the torpedo struck, the *Lusitania* lay on the bottom of the sea like a giant steel tomb. The only signs of life aboard were escaping bubbles of air: the ship's last gasp.

The sea was littered with lifeboats, rafts, floating debris, and bobbing heads. Only a few fortunates managed to survive the ordeal dry. Most people were left treading water or clinging to overturned lifeboats. Many succumbed to the cold or to exhaustion. Turner floated in the icy water for three hours before he was rescued by Master-at-Arms Williams, who pulled him onto a life raft.

Many of those who ultimately survived owe their lives to wireless operator Robert Leith, who tapped out SOS continuously for the three or four minutes before the generator died, then kept it up for several more minutes by means of emergency battery power. His message sparked a massive rescue operation that turned a dozen vessels on converging courses to a spot some ten to twelve miles off Old Head of Kinsale. For his efforts, Leith survived the sinking.

In newspaper accounts, the wealthy and the famous got most of the press coverage, but death does not distinguish between classes. The rich and the poor, the young and the old, the men and the women, the black and the white: they all die the same. Neither is heroism or cowardice an affectation of social or economic status: they are attributes of individual nature. So while the headlines shouted in capital boldface the names of the "important" people who lost their lives that day, little was written about the lesser knowns who play essential if less influential roles in the great gestalt of humanity. For survivors who lost friends, relatives, and loved ones, the suffering felt the same despite artificial class distinction.

The final death toll was staggering: 1,198 according to British court records, later revised to 1,195 in American proceedings. Queenstown was overwhelmed with the recovered bodies of the dead. They were laid out in long rows in warehouses for viewing and identification by relatives. This gruesome process continued for days.

More than a hundred corpses were never identified. These were photographed and placed in hastily-built wooden caskets. Horse-drawn hearses carried the bodies in procession along the narrow dirt streets to a cemetery outside of town. Mourners cried soulfully as local priests officiated over the burials. The dead were laid to rest in a huge mass grave later marked with a large granite rock decorated with a brass identification plaque.

A meticulous count of the sex, age, and citizenship of the victims was made for statistical analysis. Among the dead were 286 women, 72 children (both boys and girls), and 35 infants; these totaled 393, or nearly one-third of all those lost. Also among the dead were 128 Americans.

Fair Game or Foul?

The United States was shocked, England was outraged, Germany was exuberant. Officials of all three nations ran to the rule books to ascertain whether the *Lusitania* was fair game and whether causing the deaths of so many civilian passengers and crew was justifiable under the rules of civilized warfare (although the grandiloquent terms used were not "fair" and "justifiable" but "legal" and "illegal." The question of morality was not raised).

Parliamentary procedure and Roberts' Rules of Order have no place in war.

Almost without exception, the term "war" invokes aberrant conceptions of behavior in which acts otherwise offensive are deemed acceptable, in which strange notions of morality are applied, in which life is no longer held sacred, in which atrocity becomes the norm - as if a "higher" plane of reference has established new and baser standards. The sinking of the *Lusitania* attained allegorical significance.

While the general public was aghast (including some German citizens) at the appalling number of deaths, the political powers that be played games with words: literary and oratorical jousting that served no purpose against an entrenched and implacable enemy. Diplomacy ends when war begins, and resorting to demands for apologies and reparations is meaningless and wasted rhetoric while nations are still fighting, as does deliberating with a charging lion whose teeth are at your throat.

Recriminations abounded. Hearings were held in Queenstown, London, and later in New York. Many witnesses were called to testify - some say they were carefully culled from the survivors in order to sustain preconceived notions instead of to extract the whole sordid truth. No doubt this claim was true.

To be sure, a logical conclusion or sound judgment can be reached only by listening to *all* the facts and weighing *all* the evidence. When a judge selects whose testimony to hear, which evidence may be introduced, and purposely prohibits the introduction of testimony and evidence of a contradictory nature, then careful analysis yields to unwarranted assumption. That's why a court ruling is called an "opinion." Today, with records spread far and wide, it is hardly possible to know how much evidence was kept from public scrutiny. In any case, suspicions and interpretations based upon unknown and unknowable intelligence have limited value.

Lord Have Mercy, Britain Has Mersey

Presiding over the investigation conducted by the British Board of Trade was Sir John Bigham, better known as Lord Mersey. He presided over the *Titanic* probe in 1912, and the *Empress of Ireland*

30 – The *Lusitania* Controversies

hearing in 1914. Lord Mersey's expertise lay in creating scapegoats and whitewashing events, in which respects he knew no mercy and outbrushed the adventures of the eponymous Tom Sawyer. Lord Mersey handled both previous investigations with great incompetence and evident prejudice.

In the *Titanic* case, he completely exonerated Captain E.J. Smith for traveling at full speed through dense fog despite repeated warnings of ice fields ahead. The *Titanic* collided with an iceberg and went down with the loss of more than 1,500 lives. Mersey blamed the high number of fatalities on the master of the *Californian*, Captain Stanley Lord, who had nothing to do with the collision, who was more than twenty miles away at the time it occurred, and who was wise enough to heave to until morning when he could see how to steer his ship safely through the ice pack which surrounded his ship.

In 1914, he exonerated the *Empress of Ireland* for turning into the path of the *Storstad* on a fog-shrouded night in the St. Lawrence Seaway. The *Empress of Ireland* sank with the loss of more than 1,000 lives. Mersey blamed the *Storstad* for not getting out of the way of the veering *Empress of Ireland*.

Mersey was predictable, for everyone knew where his allegiance lay. He was quintessentially British. The purpose for holding the 1915 hearing was to blame Germany for sinking the *Lusitania* and to absolve British hands of all responsibility for contributory negligence. Lord Mersey saw to it that these goals were achieved - at least in the eyes of loyal British subjects and bitter American citizens.

In the *Lusitania's* case, it didn't take much whitewash to denounce the Kaiser for the disaster. Germany advertised - even glorified - the fact that a U-boat fired the fatal torpedo, and Turner told how he did all in his power to avoid it. But the issues involved went deeper than the mere allocation of blame. Most points of contention are just as sharp today as when they were first honed.

The age-old question that is still hotly debated is: did the *Lusitania* classify as a legitimate target of war? Put simply, it depends upon whose rule book was used to make the determination, and what kind or degree of infractions the opponents are willing to allow. Germany's claim that she carried ammunition was argument after the event, since Schwieger did not have that knowledge in hand when he fired the fatal torpedo.

In any case, since England and Germany each wrote their own rules and repudiated the rules of the other, the logical solution to the conundrum is both yes and no. In British eyes the torpedoing of the *Lusitania* was wrong, in German eyes it was justified. Further debate is meaningless.

THE CONSPIRACY THEORY

More important to British pride was whether the fatal meeting between liner and U-boat could have been averted, and whether the Admiralty exercised proper caution in protecting merchant shipping in light of recent U-boat activity.

Conspiracy theorists today challenge the coincidence of circumstances that placed the *Lusitania* in the path of a prowling U-boat. They would like to believe that the entire episode was a carefully contrived plot designed to incite the United States to rally for the British cause. Absent any documentation to substantiate such a notion, the assumption goes that Winston Churchill, then First Lord, proceeded beyond the pale of humanitarian regard and conspired with high-ranking officers within the Admiralty to have British patrol craft withdrawn from the *Lusitania's* predicted path and from where U-boats had been spotted and were known to be lurking, in order to expose the liner intentionally to danger. This hypothesis rests solidly on nonexistent information and on negative interpretations of facts in evidence, which together construe the "proof" of conspiracy.

The stage in this fiction was set with the dismissal of Captain Dow as the *Lusitania's* master, presumably because he was too bold a commander. His replacement was either less aggressive or he was persuaded to let his ship be exposed, and perhaps even deliberately assisted in her demise by steering into the area where U-boats were active, and by reducing speed and steering a steady course in order to make an easy target. These strong accusations were made against Captain Turner.

None of the evidence brought to light during the public hearings bore witness to these charges of conspiracy. To be sure, certain information was withheld from public view and for a very good reason: any statement of facts made public was made available to the enemy. But even today, with the keen vision of hindsight and with accessibility to those documents and testimony which were originally withheld, no incriminating evidence obtains.

Secret Admiralty dispatches revealed the weakness of British naval defense in the Irish Sea, and a knowledge of approximate U-boat movements acquired by breaking certain German codes. The plain fact of the matter was that the home waters were inadequately protected due to the shortage of available warships. This deficiency has been misconstrued by conspiracy theorists who choose to believe that the *Lusitania* was deliberately denied an escort.

To put events in perspective, bear in mind that never before had the *Lusitania* been escorted; she was considered too fast to require an escort. Every day more than two hundred merchant ships arrived in British ports, and another two hundred departed. All went unescorted. The *Lusitania* was therefore one among many of less strategic importance than an incoming tanker brimming with oil or a freighter filled with munitions.

Conspiracy theorists also contend that the cruiser *Juno* could have escorted the *Lusitania* on the last leg of her voyage, but was withdrawn from the area by secret Admiralty instructions so the liner would be unprotected against the oncoming U-boat. In reality, the *Juno* was slower than the *Lusitania* and could not have kept up with her unless the liner reduced speed. It is only coincidence that the *Juno* preceded the *Lusitania* along a similar course and docked in Queenstown. This chance occurrence was further obfuscated by the *Juno's* subsequent movements.

After word was received about the *Lusitania's* loss, the *Juno* was dispatched to pick up survivors. Shortly after leaving port she *was* recalled. By that time more than a dozen fishing boats and small commercial steamers had arrived on site so that rescue operations were progressing satisfactorily. The *Juno's* recall was based upon past dire circumstances in which U-boats submerged and lingered in the vicinity of their victims in order to pick off warships rushing to the rescue. There was no need to expose the *Juno* unnecessarily. Neither is there need to point the finger of conspiracy at an established naval protocol not understood by civilians.

THE GRILLING OF CAPTAIN TURNER

In camera proceedings with Captain Turner disclosed specific Admiralty instructions for the protection of merchant ships against U-boat attacks. Lord Mersey examined these instructions minutely, mercilessly, and ad nauseam, drilling Turner on each point over and over again, posing the same question never less than a dozen times, as if the sheer weight of discussion could alter the truth. Turner always responded simply and honestly so his testimony went unchanged through two days of intense interrogation. Mersey was either incapable of comprehending or unwilling to accept Turner's ingenuous explanations for his decision making process and consequential course of action.

Part of Mersey's difficulty stemmed from his incredible display of ignorance of all matters nautical, leading one to wonder why a person more knowledgeable in maritime affairs wasn't chosen to lead a maritime investigation. For example, after Turner explained in great detail the necessity for obtaining a fix on land in order to pinpoint the ship's position after the transatlantic crossing, this interchange occurred:

Mersey: Do you mean to say you had no idea where

you were?

Turner: Yes, I had an approximate idea, but I wanted to be sure.

Mersey: Why?

Turner: Well, my Lord, I do not navigate a ship on guesswork.

Mersey: But why did you want to go groping about to try and find where land was?

Turner: So that I could get a proper course.

Nor was this the end of the matter. Mersey kept returning to the point of precise navigation, as he kept returning to every other point he failed to grasp. He badgered Turner so harshly that one can almost hear the cruel inflections of tone that go unrecorded in a printed transcript.

Butler Aspinall, Turner's attorney, first had to give Mersey a crash course in ocean navigation so that a man who knew nothing at all about the subject could pass judgment on the actions of a man who had worked on ships at sea for more than forty years. Mersey also listened to days of testimony and contemplated for months over what he heard and subsequently read, in order to determine if the decisions Turner made on the spur of the moment were the wisest choices he could have made (without the benefit of due deliberation and foreknowledge of the outcome).

Mersey's strength lay in "Monday morning quarterbacking," a phrase which did not then exist, but one which perfectly describes the tactics in which Mersey excelled.

Ambiguous Admiralty Instructions

Rather than follow Mersey's relentless reiterations, the overall picture in contention can be brought into sharper focus by playing point-counterpoint with the various permutations of Admiralty instructions and with Turner's interpretations and attempts at compliance.

Ships should give prominent headlands a wide berth. Turner usually passed within a mile or two of Old Head of Kinsale. On this occasion he did not approach closer than ten or twelve miles. It was necessary to sight land and take bearings in order to determine his position, so he could know in which direction to proceed during fog without fear of running aground. "Wide berth" is not a precise minimum distance but a subjective term that is open to interpretation. Furthermore, this instruction contradicts another Admiralty instruction which reads: *Territorial waters should be used when possible. Remember that the enemy will never operate in sight of land if he can*

34 – The *Lusitania* Controversies

possibly avoid it.

Steer a mid-channel course. By no stretch of the imagination can the Irish Sea be considered a channel. The distance from Old Head of Kinsale to the British shore is 140 miles. Turner intended to steer a mid-channel course once he reached St. George's Channel, but that was more than 100 miles away, some six hours steaming time, from where the *Lusitania* was torpedoed.

Pass harbors at full speed. The top attainable speed of the *Lusitania* was 25 knots. With six boilers shut down her top speed was reduced to 21 knots. The morning of her loss she encountered dense fog, slowed to 15 knots, then increased speed to 18 knots for the duration. Even at that speed the *Lusitania* was the fastest merchant ship in the British fleet, so that Mersey's suggestion that the other boilers be fired up is not sustainable. In addition, maintaining full or emergency speed placed a terrible strain on the boilers, with the result that the smokestacks expelled great clouds of black dust high into the air, making the ship more visible from farther away.

Important landfalls in this area should be made after dark whenever possible. Turner explained that the 18-knot speed was calculated to place him off the Liverpool bar before dawn and when the tide was right for passage, so he could pick up a pilot without slowing down and without sitting outside the bar like a sitting duck. Mersey suggested instead that he should have maintained full speed and steamed around in circles in St. George's Channel so as to consume the extra time.

Fast steamers can considerably reduce the chance of a successful surprise submarine attack by zigzagging, that is to say, altering course at short and irregular intervals, say, 10 minutes to half an hour. Turner freely acknowledged that he did not zigzag, and he was severely castigated by Mersey for not doing so. Turner pleaded in his defense that he interpreted the instructions to mean to zigzag *after* a U-boat was sighted, as a method of escape, not as a matter of routine in an area known to be infested with U-boats. This interpretation may seem naive to us today and to militarists at the time, but we must remember that Turner was a merchant ship captain who knew not the ways of naval warfare. The most he can be accused of in this or any regard is a less than full appreciation of the submarine menace. Notwithstanding the above, the *Lusitania* actually did change course twice in the hour prior to being torpedoed. The first change was made toward land when it was first sighted, in order to determine what point of land it was (Brow Head); the second change was made away from land and was held while a fix was being made. Although these course changes were not made under the guise of zigzagging, the effect was pretty much the same.

When making principal landfall at night they should not approach

The Birth and Death of the *Lusitania* - 35

nearer than is absolutely necessary for safe navigation. This instruction contradicts itself within the same sentence. A ship cannot make landfall without first approaching it.

Finally, we come to the wireless messages received on the morning of the catastrophe: *"Submarines at Fastnet and Submarines sighted south of Cape Clear proceeding west."* Since the *Lusitania* had already passed Fastnet and Cape Clear, which are off the south coast of Ireland, these messages contained no useful information for Turner. Not so *"Submarines active in southern part of Irish Channel. Last heard of 20 miles south of Coningbeg Light Vessel."* The Coningbeg lightship guarded the entrance to St. George's Channel. If Turner had taken the course that Mersey insisted he should have taken, he would have steered the *Lusitania* through an area that the latest Admiralty intelligence and decrypted intercepts considered the riskiest.

Nor is the Admiralty necessarily to blame for issuing paradoxical warnings. The advices to avoid headlands and to steer a mid-channel course were given in general, whereas the advices about U-boat activity in the south part of Ireland and in the area twenty miles south of the Coningbeg lightship resulted from the latest sightings and decrypts, and, as such, were local updates valid for the time of issuance. Thus the latter supersedes the former. Admiralty advices were issued with the intention of supplying masters with the wisest counsel and latest intelligence so they could make informed judgments on the best way to proceed, not to tell masters how to run their ships. Furthermore, the Admiralty did not intend its instructions to conflict with Board of Trade regulations for prudent navigation.

BACK TO CONSPIRACY

Conspiracy theorists take exception to the fact that the U-boat updates of May 6-7 were not directed to the *Lusitania* specifically by name but to merchant shipping in general, thus softening the impact the warnings should have made on Turner. They also claim that Turner was purposely herded into the path of the *U-20* by these warnings. These subjective evaluations of the same warnings are mutually exclusive.

The torpedoing of the *Lusitania* was the result of fate, destiny, chance, or coincidence - but most certainly not conspiracy. That she steamed into the path of a lurking U-boat whose captain was on the prowl for targets of opportunity is no more suspicious in this instance than in the thousands of other instances in which merchant ships met with similar demise. Such are the fortunes and misfortunes of war.

At the time of the *Lusitania* disaster Churchill was in France conducting secret negotiations. After his return to England he took great pains to arouse public fervor over this most recent German atrocity,

36 - The Lusitania Controversies

and made sure to point out that 128 Americans died in the sinking, hoping to involve the United States in the struggle against the Kaiser's bid for world domination. This blatant attempt to exploit the situation after the fact was shrewd but justifiable. One cannot argue backward from that point to extrapolate a precondition of conspiracy, or to prove that Churchill deliberately plotted the torpedoing of the *Lusitania*. In order to do so he would have needed German cooperation, which he certainly did not have.

It is interesting to note that this same claim of conspiracy was made against Churchill again in 1939 when a German U-boat torpedoed and sank the liner *Athenia* with great loss of life.

LAND OF THE FREE, HOME OF THE TIMID

Despite Churchill's stratagem and the abomination of the loss of so many innocent lives, the U.S., although vocally remonstrative, remained resolutely neutral in action. Those today who believe that the sinking of the *Lusitania* provoked the U.S. to declare war against the Central Powers need to be reminded that such a declaration was not proclaimed until nearly two years after the event, on April 6, 1917 - a slow fuse by any measure.

This is not to imply that American citizens and their leaders were not incensed by the horrible deed, only that they were not incensed enough to do anything about it. Talk is cheap: the tool of the weak-willed and cowardly. A browse through contemporary newspapers divulges countless editorials and impassioned pleas that condemned the act as heinous, and demanded a firm apology for the "crime against civilization and humanity." Vituperation may be psychologically satisfying to some, but rhetoric is pointless if it does not lead to a remedy or adequate compensation. In the latter regard, none of the lives lost in the sinking could be brought back from the dead. What remedy was there to seek?

President Woodrow Wilson sought "strict accountability," a pompous piece of nonsense if there ever was one. Worse yet was his characterization of the torpedoing of the *Lusitania* as "so absolutely contrary to the rules, the practices and the spirit of modern warfare." I reiterate, rules are for games like poker and Parcheesi, not for international acts of violence committed in pursuit of world domination. Offensive war is inhumane; defense against it is imperative. The mention of "spirit" suggests festivity rather than bloodshed.

Wilson's effete grandiloquence was a far cry from his predecessor's political policy: "Speak softly, but carry a big stick." If Teddy Roosevelt had been running the country instead of traipsing through South American jungles and paddling a dugout canoe up the unknown River of Doubt and down the treacherous Amazon River, he would no doubt

The Birth and Death of the *Lusitania* - 37

have led the United States in the charge across the Atlantic Ocean and up the hill of the German aggressors.

Not so Wilson. He spoke softly but the stick he carried was limp - a mere willow that easily bent in the breeze. Under Wilson's administration, Roosevelt's Great White Fleet had become a Pretty Pink Elephant, due largely to the ineptitude of Josephus Daniels, the Secretary of the Navy whose criminal mismanagement of naval forces had yet to be fully realized. Add to this the implacable pacifism of William Jennings Bryan, the Secretary of State, and the recipe for American isolationism was complete. The severing of diplomatic relations with Germany was Wilson's biggest threat. He wanted to maintain "peace at any price."

Nowhere in all the editorial protests, in senatorial and congressional speeches, in diplomatic dispatches, or in presidential oratory, was there any suggestion that the United States should go to war: empty words all. Nor were there any rallying cries. The *Lusitania* crisis kindled no inflammatory slogans like "Remember the Alamo" or "Remember the *Maine*." The *Battle Hymn of the Republic* was a forsaken anthem.

Teddy Roosevelt stood practically alone when he proclaimed, "There are things worse than war."

The majority of American citizens were content to make faces and shout denunciations from across the barricade of the broad Atlantic, complacent in their noninvolvement and willing to let England and France fight the battle for world freedom.

Riding on a tide of arrogance, Germany added insult to injury by decorating Schwieger for his daring exploit. The tide soon turned, however, as world opinion mounted against Germany's tactless insolence. It is difficult for people today to feel the emotional tension in the geopolitical arena because we enjoy the benefit of retrospect; we're not blind like they were then to the possible outcome of events.

Bryan resigned in protest during the *Lusitania* crisis because he feared a declaration of war. He was replaced by Robert Lansing whose views were clearly anti-German.

Germany was becoming desperately afraid that England would convince the U.S. to join the fight. American allegiance and violation of neutrality had clearly been established by the sale of munitions to England - munitions which were precluded from sale to Germany, a fact that Germany was quick to mention before offering diplomatic concessions. Germany remained firm on the point that the *Lusitania* classified as a legitimate target that had been given fair warning in American newspapers, and made no apologies for the untimely end of those Americans who chose to take passage on a British vessel through the zone of war. However, Germany presented a tentative conciliation intended to placate President Wilson and indignant

38 - The *Lusitania* Controversies

American citizens: the cessation of unrestricted submarine warfare.

A country or person unwilling to fight must accept the concessions offered.

Short-Lived Concession

On June 6, 1915, German U-boat commanders were ordered to abide by the rules of the Hague Convention. It made little difference. Neutral vessels were nevertheless sent to the bottom in droves. On August 19 the White Star liner *Arabic* was torpedoed by the *U-24*, with the loss of over forty lives, some of whom were American. Thus Germany's offer of settlement was in constant violation: a paper policy at best since the Kaiser knew full well that he could not win the war "fairly." Further protests from England and America merely padded diplomatic pouches.

It is interesting to note that the U.S. protested vigorously the British blockading system and the mining of the North Sea because it constituted a violation of neutral rights and freedom of the seas. The U.S. had already fought England over that very same issue in the War of 1812. As a result of the *Lusitania* incident, relations deteriorated between the U.S. and Germany, and strengthened between the U.S. and England. Ironically, once the U.S. entered the war on the side of the Allies, it extended the North Sea mine barrage that it had once protested, and deployed more mines than England ever had. Such are the changing values in world politics.

Atrocity Commemorated

To rub more salt into British wounds, German metalworker Karl Goetz minted a medal commemorating the *Lusitania's* loss. The medal was not unique. Nearly six hundred commemorative medals of various kinds were struck by private German mints during the war. What brought Goetz's metallic lampoon into dispute were the lack of respect that was shown for the dead, and an erroneous date.

The medal depicted passengers lined up like cattle to buy tickets from Death personified as a skeleton, with the blessing of Cunard, and a ship bristling with guns and ammunition. The medal was not widely circulated and went barely noticed in Germany. But England counterfeited it by the tens of thousands and distributed it to an aroused civilian populace, to further fan the flames of fervor.

May 5 was stamped on Goetz's first lot as the date of the *Lusitania's* loss, instead of May 7. This has given rise to speculation that the medal was struck in advance, in bold anticipation of the accomplishment of so difficult a task, which accordingly must have been premeditated - more twisted evidence for conspiracy theorists. The true explanation was simpler: a typographical error. The real

The Birth and Death of the *Lusitania* - 39

import of Goetz's medal was that it kept the tragedy of the *Lusitania* alive in the minds of the people.

The German Abomination

Reaction to the *Lusitania* atrocity and association of it today stand way out of proportion to the more barbaric outrages committed by the Kaiser's army coevally. Consider the barbarity of the German invasion of Belgium, the bombardment of the cities, and the ruthless rape, pillage, and indescribable mayhem that was the norm for the advancing Hunnish horde.

Entire towns were destroyed and the adjacent lands laid waste. Churches, cathedrals, and ancient libraries were incinerated, monuments were blown apart, historic medieval buildings and architectural works of art were demolished. Thousands of houses were torched and their inhabitants burned alive inside them; those who tried to flee the flames were gunned down by German rifle fire. Hundreds of thousands of Belgians were forced to flee to Holland, some walking the whole way without food or belongings. Survivors who stayed behind were indiscriminately bayoneted by invading German soldiers.

In addition to the tens of thousands of civilians who were killed by shell fire and bombardment, thousands more were summarily executed by hanging, by stabbing, by dragging, by battering, by burning, by blasting, and by shooting. Among the dead were a large proportion of old men, women, and children. Theft and the mutilation of corpses were the least despicable of the carnage and crime. Germany justified such wholesale annihilation on the grounds of military necessity.

These descriptions may sound tame compared to later Nazi excesses, but they go a long way toward demonstrating that the torpedoing of the *Lusitania*, as nefarious as it was, stood as a minor episode in relation to the war as a whole. More than eight *million* human beings were slain as a result of German agression in World War One. Against the pall of such savage extravegence, the nearly 1,200 people who went down with the *Lusitania* comprise but a negligible part of the whole.

Ultimately, the question of whether Germany had the right to sink the *Lusitania* can be answered by simple analogy. If Germany had the right to enslave the world, then it had the right to slaughter those who interfered with its goal.

Life after Death

Ships die not unlike people. Some die bravely, some die violently, some die accidentally, some die anonymously, some die notoriously. And some achieve immortality in the mind of man.

The saga of the *Lusitania* will never die. It will live on forever.

Part 2
Fame Everlasting

THE *LUSITANIA* MYTH

All of the actual and insinuated events related in the previous pages have contributed to the creation and entrenchment of a mythos which has been embellished by makers of epic. Sadly, mention of the *Lusitania* often brings to mind a thrilling web of intrigue instead of the ultimate horror of war and the senseless prosecution of death.

Those who condemn the Admiralty for faulty action or an apathetic attitude perceive the *Lusitania* in microcosm, as an isolated event. They fail to see the broader perspective of the Grand Fleet's responsibilities, the most important of which was to prevent the Imperial German Fleet from putting troops ashore on the British coast and shelling cities and towns, an invasion that would have had a devastating effect on the British crusade and that would have caused casualties too numerous to tally. This is to say nothing about the defense of the English Channel against German minelayers and patrol boats, offensive operations in the Black Sea and in the Mediterranean in support of the stupendous land battles then raging, and the worldwide search for German surface raiders that were disrupting merchant shipping across the globe. The Admiralty's forces were seriously overextended, and its attention was diverted by more conspicuous military demands. By contrast, the imminent arrival of the *Lusitania* was little more than a distraction.

Against that backdrop, it must be emphasized that the *Lusitania* was only one of thousands of ships that were sunk during the war to end all wars. In his six-volume history of the Great War, *The World Crisis*, of more than 3,000 pages of text Winston Churchill devoted only two to the *Lusitania*. Such was her significance relative to the greater insanities in progress.

Yet the *Lusitania* is remembered far better than more degenerate contemporary atrocities that occurred in World War One: battles in the which the loss of life was truly enormous. To cite a few:

- First Battle of Ypres, in which losses for the month-long struggle totaled 100,000 troops.
- Second Battle of Ypres, in which the Germans launched the first large-scale poison gas attack and killed 60,000 Allied soldiers (and permanently disabled countless others).

Fame Everlasting – 41

- The Gallipoli Campaign, which is more famous for the stealthy British retreat than for the loss of 570,000 Allied and Axis troops.
- First Battle of the Marne, in which total losses amounted to 550,000 men.
- First Battle of the Somme, in which 19,000 British soldiers were killed in a single day. The total losses for the summer campaign exceeded one and a quarter *million* troops.
- Second Battle of the Somme, in which half a million soldiers were killed in the field.

Today, mention of these name places is more likely to bring acknowledgment of the strategic importance of the battle sites than of the number of casualties that occurred there.

NOT TO LANGUISH IN OBSCURITY

Fame can be fleeting or everlasting.

An object of fleeting fame soon fades into obscurity.

Fame everlasting comes in various degrees: an item may be nothing more than a footnote in a history book, or it may assume prominence that is way out of proportion to its original merit.

Fleeting fame may later be revived and surpass in notoriety items of everlasting fame.

This discourse is not intended as an exercise in logic. More often than not, fame of any sort is based upon no rational premise. Some events of global significance never make the grade and are forgotten forever, while others that are distinctly trivial assume legendary status in the social consciousness.

Witness the absurd consent of greatness given to professional game-players who drop balls into hoops or into cups in the grass. Or entertainers who write biographies about how they furthered their own careers. Or cartoon characters that become household names. A sports figure with a sprained ankle may dominate the headlines while the starvation of human beings in a third world nation is relegated to the back pages almost as an afterthought. Thus do incidents of minor note achieve a perception of sympathy that is beyond their due.

Fame cannot be associated automatically with fundamental importance. Its relevance must be placed against a much larger tapestry.

The very mention of shipwreck has always touched a chord in the public psyche: an evocation of suffering that is shared by all and that endures more than far worse disasters and natural catastrophes. Fires, floods, famine, earthquakes, and volcanic eruptions - most come and go as quickly as the duration of the events. The emotion evoked by such tragedies soon wanes.

42 - The Lusitania Controversies

What is it about shipwreck that gives it perpetual recollection? And why does one wreck achieve immortality while another is dismissed with a shrug?

Is it misfortune? Human drama? Heroic deeds? Cowardice? Culpability? Number of fatalities? Avoidable action? Or is it cultural foible, as indecipherable as the raging popularity of a television commercial featuring dancing raisins?

No doubt people harbor deep-seated emotions which reach out to the victims of shipwreck, empathizing, for example, with the helplessness of their plight, the agony of despair, the fear of impending doom, the anguish of facing an undeserving fate. In some ineffable manner, the innocence of the victims preys upon the soul of social conscience.

Not to be ignored, however, is the fact that a shipwreck is a quantifiable event: it occupies a precise moment in time, it incorporates direct cause and effect, and - what may have major significance - it already has a name. A word, a slogan, a label - all are succinct designations around which a singular idea can be formed and fixated.

Furthermore, whereas ruined cities are later rebuilt, and ancient battlegrounds become fields for crops, and scenes of disaster are cosmetically covered, shipwrecks often lie fallow on the bottom: a hidden gestalt awaiting symbolic resurrection.

But why is one wreck found distinguishable while another is ignored? To a large extent the answer is found in the publicity it receives. After the initial furor dies down, after accusations have been made and upheld, after the guilty have been pilloried and punished, after the media moves on to the next event, closure has been duly consummated. But if this process is violated, recriminations can fester like an open sore. Continued coverage of an unfulfilled event can create more of a story than existed in the first place.

Without a doubt the most well-known shipwreck in modern times is the White Star liner *Titanic*. Hearings, court proceedings, investigations, recriminations, journalistic rhetoric - all failed to close the event with any degree of finality. Yet even the great *Titanic* tragedy sank into a sea of silent nepenthe, and was largely dismissed by later generations as a minor episode in the history of transatlantic commerce. With all the current hoopla and media attention that has been focused on the *Titanic* in recent years, it hardly seems possible that the name was not always on the tip of everyone's tongue. Yet for forty years the *Titanic* enjoyed relative obscurity, merely one shipwreck among a multitude mentioned in passing, her name appearing only sporadically in publications devoted to nautical trivia - until the dramatic story was revived by Hollywood in 1953. Riding the tide of *Titanic* movie popularity came Walter Lord's book, *A Night to Remember*. The snowball has been growing ever since.

Less well remembered (or almost forgotten) are the losses of such superliners as the *Arabic*, *Athenia*, and *Britannic* (the *Titanic's* sister ship), to mention a few. Granted, the number of fatalities in those cases was but a fraction of that of the *Titanic*. But how many recall the *Empress of Ireland*, which was sunk by collision in the St. Lawrence Seaway in 1914 and was attended by more than 1,000 deaths? Or the excursion steamer *General Slocum*, which burned in New York's East River in 1904 with the loss of over 1,000 people, almost exclusively women and children? Or the World War Two liners *Leopoldville*, *Steuben*, and *Wilhelm Gustloff*, all of which were torpedoed? More than 800 people went down with the *Leopoldville*. Over 3,000 died in the sinking of the *Steuben*. When the *Wilhelm Gustloff* sank in forty-two minutes, more than 5,000 refugees found slender solace in a watery grave. Why haven't they achieved prominence?

The list goes on and on. Tens of thousands of shipwrecks have occurred in the nineteenth and twentieth centuries alone. Perhaps hundreds of thousands have occurred since the beginning of recorded history. Only a fraction are known today. Of those known, only a small percentage come to mind when the populace is polled. And of those that are recalled, the majority receive a share of renown that is disproportionate to their significance.

Consider these recent Great Lakes shipwrecks whose circumstances were similar. The *Carl D. Bradley* sank in 1953. The *Cedarville* was lost in 1965. The *Daniel J. Morrell* went down in 1966. The *Edmund Fitzgerald* disappeared in 1971. Hardly any but local historians and wreck-divers know of the *Carl D. Bradley*, the *Cedarville*, and the *Daniel J. Morrell*, but nearly everyone is familiar with the *Edmund Fitzgerald* - simply because Gordon Lightfoot wrote a song about it. On such an air of ephemera does notoriety depend.

Do not confuse popularity with significance.

Publicity, exposure, and advertising can make people and products known. By similar means an unremarkable shipwreck can be launched into the limelight.

The sinking of the *Lusitania* has not assumed *Titanic* proportions. No shipwreck ever will. Yet there is no doubt that the *Lucy* occupies a permanent place in the annals of sordid events, and justifiably so. All the elements were there: true tragedy with respect to the incredible number of fatalities (many of whom were women, children, and babes in arms), the death of prominent civilians, the relevance of passengers who were citizens of neutral nations, the hint of indiscretion in the transport of munitions, the argument over the definition of a legitimate target of war, the question of the legality of unrestricted submarine warfare, and the incarnation of atrocity in the lust for world domination

Controversy sells. Conspiracy sells better.

44 - The *Lusitania* Controversies

THE CREATION OF A MYTH

At the time of her sinking, the *Lusitania* was front page news. For months afterward, while pusillanimous President Wilson dawdled over diplomatic communiqués that were blatantly sweetened with noncommittal grandiloquence, flames of resentment and indignation swept through England and America: flames that were fanned by the press. In America, anti-German sentiment rose to a fever pitch, then gradually faded as the public began to realize that a too-aggressive attitude could lead to war. Most Americans who demanded reparations for Germany's hostile actions were not willing to join the Allied nations in battle. America continued to hide behind the security of isolationism.

Controversy abounded. The obsession with conspiracy did not occur until later. Years later. Decades later. Half a century later.

Meanwhile, the war to end all wars became more vicious and perverted in its conduct: the senseless killing of civilians, the bombardment of ancient cities, and the worst wartime depravity prior to the application of napalm: poison gas. Against such horrors the sinking of the *Lusitania* became but a straw on the back of the camel that plodded reluctantly toward American declaration. The torpedoing that was once perceived as a lone aberration of civilized warfare, turned out to be merely one in a long list of more outrageous acts of cruelty. By the time America joined the effort to fight for freedom in the world, the *Lusitania* was but a minor incident among greater German atrocities. Her name did not even become a battle cry.

What kept the *Lusitania* alive in the mind of man was publicity, else she might have joined the ranks of other vanquished shipwrecks such as the *Arctic*, the *Eastland*, and the *Mont Blanc*.

AN EARLY PUBLICATION HISTORY

It would be practically impossible to assemble a bibliography of every mention made of the *Lucy* in books, articles, and newspaper columns. Nor does mere citation necessarily help to build a myth. More often than not, the *Lusitania* was mentioned merely in passing, more a matter of shipwreck name dropping than adding new material.

What has helped the most in maintaining the *Lusitania's* image is the publication of books there were devoted strictly to the sinking. In the present volume I have purposely avoided repeating information that has already been recorded in earlier works. An annotated recap of previously published books may be of interest to the reader who wants to pursue the issues further.

1915, *Horrors and Atrocities of the Great War, Including the Tragic*

Fame Everlasting - 45

Destruction of the Lusitania, by Logan Marshall. The title tells it all. Logan Marshall assembled memorial editions on the *Titanic*, *Empress of Ireland*, and other hyped-up catastrophes such as fires, floods, and earthquakes. All these books were widely distributed very soon after the events which they portrayed, in order to catch the tide of glamour before it began to ebb.

One third of the book focuses on the *Lusitania*, the rest recounts other German barbarities. The text largely reprints newspaper copy and editorials, with some original material provided by the author, along with special chapters contributed by three other writers. The book is justifiably anti-German and written in a style whose superlative purple prose is exceeded only by the utter horror of the events they recapitulate. The many quotes provide a compelling insight into the American perception of the war in Europe. A valuable sign of the times.

1915, *The Tragedy of the Lusitania*, by Captain Frederick D. Ellis. This is a compilation of quotes gleaned from contemporary newspaper accounts, threaded together with narrative that was written by the author. It is similar in format to the best-selling books that were written by Logan Marshall.

One might suppose that a job that was rushed would be insipid as a result, yet Ellis's contribution contains a surprising amount of depth. The text is highly readable and graphically descriptive, with perhaps too much of a tendency to suffer from excessive cloying.

The overall flavor is one of anti-German sentiment, to be expected in light of contemporary attitudes and the unavoidably harsh truth of German brutality. Yet the book as a whole is not as exploitative as it could have been. Editorials of both pro and con propaganda are included, providing somewhat of a balance, although opinions from people of German descent - Roosevelt's hyphenated Americans - are clearly in the minority.

The connective passages stress ad nauseam the death and suffering of women and children - a common attribute of contemporary disaster tales. Very little emphasis is placed on rich and famous passengers. Most of the quotes originated from second and third class passengers. Definitely worth reading.

1915, *The Lusitania's Last Voyage*, by Charles Lauriat. This is a first-person account that was written by a survivor. The slender volume consists of two long letters which Lauriat wrote to his mother, the combined length being not much longer than a pair of magazine articles. Lauriat refrains from resorting to hyperbole. His narrative is strictly factual and surprisingly underwritten. The reader almost wishes for a bit more pain and suffering. Interesting and worth reading.

1916, *The Lusitania Case*, collected by C.L. Droste. This book differs from its predecessors in that it does not detail the circumstances of the sinking and the drowning of the passengers and crew, but is devoted instead to the legal ramifications and the moral issues within the context of "civilized" warfare - an oxymoron if there ever was one. The book is a collection of opinions and editorials taken from newspapers published on both sides of the Atlantic, sworn affidavits, official releases, bureaucratic comments, boards of investigation, and German government sources.

In some cases the quoted comments lack validity because they were made by people with little or no expertise in the matters under discussion, but overall the text is based on material that was provided by knowledgeable consultants. The book makes its points without resorting to exaggeration. Because most of the references are diplomatic in nature, and lacking the overblown emotionalism of Marshall's and Ellis's books, the effect is both thought provoking and intellectually stimulating. A small gem.

1918, *German Submarine Warfare: a Study of its Methods and Spirit, Including the Crime of the Lusitania*, by Wesley Frost. This is a knowledgeable, intellectual perspective from the pen of the U.S. consul in Queenstown at the time the *Lusitania* was torpedoed. The first two-thirds of the book describes U-boats and their methodology of attack, the legality of unrestricted submarine warfare in relation to humanitarianism and international politics, and some of the submarine attacks on merchantmen, with particular attention paid to the German policy of leaving survivors to fend for themselves when they were hundreds of miles from land and were left with no hope of rescue. The overtones are decidedly anti-German.

One of Frost's responsibilities as consul was to interview survivors of U-boat attacks who were landed at Queenstown, then a major port in southern Ireland. He conducted his official duties with incredible vigor, maintaining meticulous records of U-boat activity and writing down the statements of each and every survivor. His insights are invaluable.

The final one-third of the book is devoted to the *Lusitania*. Frost reports on rescue measures and the immediate aftermath of the sinking: obtaining lodging, food, clothing, and loans for survivors; notifying friends and relatives of the dead; photographing and identifying bodies; and authorizing the construction of hundreds of coffins and then attending to the burials. A fascinating homage to ratiocination.

These five books appeared in the first three years after the *Lusitania's* sinking, when public indictment of the atrocity continued to exert some influence over American opinion.

In England the *Lusitania* was but one of a thousand U-boat victims, and the civilian casualties were nothing compared to the carnage occurring on the western front. The British shrugged it off and concentrated on fighting.

After the U.S. entered the war, in April 1917, the horrors of mass death and destruction occupied so much of the collective consciousness that outrage over the torpedoing of the *Lusitania* assumed a lower place in the hierarchy of more terrible depredations.

The *Lusitania* was not forgotten, but neither had it captured the imagination. To be sure, the name became a household word like *The Wreck of the Hesperus*: that evocative poem by Henry Wadsworth Longfellow which did for the *Hesperus* what Gordon Lightfoot's ballad did for the *Edmund Fitzgerald*. But the *Lucy* persisted more in the way of a statistic than as a posthumous embodiment of horror, pain, and suffering. The *Lusitania* was history - but a minor event in the war to end all wars.

Lusitania and the Silent Screen

In addition to the published references noted above, the *Lusitania* was the subject of a newfangled presentation medium which was then in its infancy: the movies. *The Sinking of the Lusitania* was the brainchild of American cartoonist Winsor McCay, a pioneer in animated films. McCay was famous for two pre-war cartoons, *Little Nemo* (1911) and *Gertie, the Trained Dinosaur* (1914): "shorts" that were shown to audiences in theaters to the accompaniment of piano music. These cartoons probably added diversion to live-action plays rather than preluded longer motion pictures as did cartoons of later-day popularity.

The Sinking of the Lusitania was released in 1917 or 1918 (sources vary). It is considered to be the first feature-length animated film. I have seen *Gertie, the Trained Dinosaur*, but whether prints still exist of *The Sinking of the Lusitania* I don't know. Perhaps an original celluloid still exists in a dusty can in a film archive, awaiting rediscovery.

Lusitania on Stage

In 1926, the still controversial sinking was acted out on the stage. It first appeared on January 18 in Darmstadt, Germany. The play was both "hissed and applauded." A contemporary review reads:

48 – The *Lusitania* Controversies

"Nationalists and Modernists gave a mixed reception last night to the premiere of Alfred Doeblin's "Lusitania," which dramatically depicts the torpedoing of the liner, experiences of some of the characters on the bottom of the sea, and finally their rescue.

"The Nationalists considered the subject ill-chosen, especially as many of the victims in their death agonies are made to swear vengeance against those who sank the liner. Hisses, catcalls and stampings of feet were evidence of this disapprobation.

"The Modernists acclaimed the piece as analyzing in a new and daring way the feelings and thoughts of people in moments of greatest danger when egoism is revealed in its crassest form."

Then came a hiatus of thirty years before the popularization of shipwrecks became a craze, and the *Lusitania* saga was again revealed to the public - the public of a different generation: one that was somewhat inured to the horrors of Germany's previous bid for world domination by more recent German atrocities, such as wholesale slaughter and the pogrom against the Jews. Compared to these, the sinking of the *Lusitania* and the loss of nearly 1,200 people paled to insignificance.

REPARATIONS FOR LOSS

In the decade following the Great War, the *Lusitania* was kept in the American limelight largely by the filing of claims against Germany for the compensation of losses incurred in the sinking. It seems patently absurd that 128 American citizens should receive such special status, while the rights of the millions of other casualties of war were for the most part ignored. Yet the Treaty of Versailles contained provisions that permitted *Lusitania* victims to sue for injuries and to recoup their personal losses, and enabled heirs to make claims against loss of life.

Germany voluntarily signed the Treaty of Versailles at Berlin on August 21, 1921. Not that Germany had any choice in the matter. If "to the victor go the spoils," then to the vanquished goes the responsibility of making reparations to its victims and their children.

The United States established the Mixed Claims Commission to deal with claims resulting from war losses. A Senate investigation enumerated American claims with respect to the *Lusitania* thus:

Loss of life	$14,215,117
Personal injuries	967,812
Cargo losses	34,064
Loss of personal effects	277,215
Total	$15,494,208

A complicated equation was created in order to establish fair awards. "In death cases the basis of damages is not the value of life lost or the loss sustained by the estate of the decedent, but the losses to claimants resulting from his death in so far as such losses are susceptible of being measured by pecuniary standards." In plain language, "where an heir suffers no pecuniary loss through the death of a relative or was not dependent for support upon that relative who went down in the *Lusitania* there shall be no award."

The detail was quite picayune. "Bearing in mind that we are not concerned with any problems involving the punishment of a wrongdoer, but only with the naked question of fixing the amount which will compensate for the wrong done, our formula expressed in general terms is:

"Estimate the amounts (a) which the decedent, had he not been killed, would probably have contributed to the claimant, add thereto (b) the pecuniary value to such claimant of the deceased's personal services in claimant's care, education or supervision, and also add (c) reasonable compensation for such mental suffering or shock, if any, caused by the violent severing of family ties as claimant may actually have sustained by reason of such death.

"The sum of these estimates, reduced to its present case value, will generally represent the loss sustained by claimants. . . . Mental suffering to form a basis of recovery must be real and actual, rather than purely sentimental and vague."

Moreover, "Interest at 5 per cent. was granted in all cases, dating in some of them from May 7, 1915, and in others from Nov. 1, 1923."

Even in death there are class distinctions. Elizabeth Jane Bremmer was awarded two claims aggregating $600. May Davies Hopkins, whose husband was the president of the Newport News Shipbuilding and Drydock Company, was allowed $130,000. Ironically, the wife and children of millionaire Alfred Gwynne Vanderbilt were awarded nothing; their claim was disallowed on the argument that the proceeds of Vanderbilt's estate generated more income for his heirs than they were receiving prior to his death.

The first awards were not arbitrated until February 1924 - nearly nine years after the sinking. Awards were issued sporadically for the next three years. By that time, the amount of awards that were not discredited reached some $180 million. This was not just for *Lusitania* claims, but for all American claims from individuals as well as from businesses, including oil companies, insurance companies, and owners of cargo that was lost due to enemy action.

Germany did not have to put up any money in 1927. Instead, in a scheme which the word "absurd" fails to adequately describe, the U.S. government lent the money to Germany, expecting Germany to pay back the loan with interest over a term of several years.

MEMORIAL FOR THE DEAD

Another way of keeping the *Lusitania* alive in the mind of the public was by the dedication of a memorial statue in the Irish town of Cobh (formerly Queenstown), where the victims of the sinking were landed. A mass grave in the local cemetery was already marked with a plaque in memory of the dead. Thus the town was the obvious and appropriate location for the erection of a statue in further honor.

Fund raising for the project was organized by a committee of Americans which was chaired by William H. Vanderbilt. Authorization for acceptance of the bronze and marble statue was announced by the Cobh town council in 1927, twelve years after the event. Irish-American artist Jerome O'Connor was chosen to design the sculpture. His design was approved by "the Cunard Steamship Company's representative in the Irish Free State."

The project languished for next six years, probably because of politics. Not until 1933 did the town council provide a site for the monument. By then, O'Connor could not be found; he lived in the United States. The project was on the verge of cancellation when the sculptor was finally located.

O'Connor said, "The memorial would be a gift from one friendly nation to another and a peace offering in memory of all those innocent persons who lost their lives by the sinking of the *Lusitania* off the South Irish coast during the World War."

The statue was finally constructed (of Irish materials) and erected in 1935. It still stands today in the town square.

THE LEAVITT-LUSITANIA SALVAGE COMPANY

It may come as no surprise that schemes to salvage the mammoth liner were being considered by commercial salvors - not to raise the wreck but to recover some of "her valuables worth several thousands of pounds." Notice of intent to salvage was first publicized in 1919.

In January 1920 it was reported that "efforts to raise some of the treasure from the *Lusitania* will be made early this year. Engineers and divers who have been prospecting about the sunken vessel believe that they can at least get some thousands of pounds' worth of valuables from the big Cunarder, but it will be impossible, according to the experts, to raise the steamer or any of her cargo owing to the great depth of the water in which she is lying."

Nothing about salvaging the *Lucy* was heard again until 1922. Already a myth was in the making. Somehow, the thousands of pounds of valuables mentioned in 1919 and 1920 had escalated in value to millions of dollars in gold - a phenomenal and unprecedented rate of currency exchange for a mere two years passage. Human belief systems being what they are, many people had no difficulty lending credence to the suddenly inflated claims, despite cargo manifests that listed no such quantity of gold.

The Lusitania Salvage Company was formed in Philadelphia. The company chartered the steamship *Blakely* "for the salvaging of the *Lusitania* and other ships that have gone to the bottom of the ocean in the last few years." This comment makes it seem as if the *Lusitania's* name was mentioned because of her fame, not her fortune, as a way to obtain investors. Company representatives went on to state that "a single passenger is said to have deposited $75,000 with the purser of the *Lusitania*, while Mme. Antoine de Page, wife of the medical director of the Belgian Red Cross, is reported to have carried more than $100,000." Note the use of such hearsay phrases as "is said to" and "is reported to." No claim to the veracity of these statements was made.

The company's announcement of specific intent initiated an immediate storm of protest.

The British Admiralty stated that "as the British Government had undertaken the insurance of merchant vessels during the war, any salvage would belong to the Government."

According to the marine department of the British Board of Trade, "nothing is known of any proposal to salvage the *Lusitania*. The question of the right of the Government to do so depended upon the position of the wreck. If it lay within the three-mile limit the Government would possibly have a lien upon it."

Despite these comments, the secretary of the Already Salvage and Towage Syndicate stated that "the *Lusitania* was demised to my company for three years and the company, so soon as circumstances permit, will proceed with the salvaging of the ship and its contents. No other company or person has any right to undertake the work during the continuance of this demise."

It is odd that no one in the government knew anything about such a demise.

Nor was the Lusitania Salvage Company dissuaded from its quest by these contrary claims of priority. The *Blakely* was outfitted with nitroglycerin, a thirty-ton boom, and special diving suits invented by the company's president, Benjamin Leavitt. In a contradictory mix of engineering knowledge and general ignorance, Leavitt explained that "these suits, which are metal forts with jointed limbs and heavy glass portholes for sight, are equipped with tanks to supply oxygen and with caustic soda to take up the carbon dioxide. In an ordinary suit the diver would be smashed at that depth as if a building had fallen on him. If he escaped crushing, he would probably be killed by the 'bends,' a disease induced by pressure, as he was lifted to the surface."

While it is true that scientists understood very little at the time about the bends, other than the need for slow decompression, Leavitt's declaration that a diver in ordinary hard-hat dress would be smashed at that depth is nonsense. Leavitt was obviously aware of the actual truth when he denounced a claim made in 1918 by one H. Ensor before the Engineering and Scientific Association of Ireland, to the effect that the *Lusitania* "had ceased to be a ship" because "she was subjected to the enormous pressure of 140 pounds per square inch."

Other marine engineers agreed, arguing that "the terrific pressure of the water at the depth of 275 feet or thereabouts to which the *Lusitania* sunk, must have crushed her like an egg-shell."

Leavitt refuted Ensor's allegation by extolling that "a bottle air will be smashed when it is lowered ten feet below the surface. A bottle of water may be lowered 10,000 feet without being crushed. There is no doubt in the world but that the *Lusitania* exists today exactly as it did when it sunk." Leavitt held that "water, forcing its way into the *Lusitania* as it sunk, kept the pressure outward equal to the pressure inward, except in a few airtight compartments."

Leavitt is indisputably correct, but by the same token neither is a diver crushed as long as the air he breathes is compressed to that of the ambient pressure of the water. In fact, marine engineers willingly conceded that "while divers have explored at the depth, none has been able to work so deep."

With stereotypic treasure salvor hyperbole, the Lusitania Salvage Company expressed firm optimism that divers "encased in a rigid shell of metal and hard rubber" could use nitroglycerin to "cut through three decks of the *Lusitania* and raise to the surface $5,000,000 in gold from the strong room and the purser's safe, said to contain $1,000,000 in gold and jewelry." The boom would then be used to "lift up the safe and the boxed gold" through the holes in the decks.

When confronted with opposing claims, Leavitt said that he would take his chances in the International Salvage Court. "I believe that we will get by far the greater part of the value of the treasure. Any salvage court would allow it to us. We have made no arrangement in advance with owners, shipping companies, insurance companies or the British admiralty, and we do not need to, because a salvage company has the absolute right to proceed on its own initiative in the case of a vessel sunk so deep and sunk so far from shore. Our rights will be taken care of thereafter in the courts."

Leavitt commented, "If the weather is pleasant, we will cut the treasure out of her in three weeks. Storms or bad weather may delay it, but twenty-one days of good weather will see the job completed."

The expedition was expected to cost $160,000, "as against the prospect of the lion's share of $5,000,000." Leavitt asserted that "some big financiers have sought to interest themselves in it during the last few days, but it is too late. I was offered $550,000 for the proposition as it stood, but refused. The investors who have put up the money to back the expedition are all middle-class people of moderate means."

This last statement seems to have been pure bravado, for the next we hear of the Lusitania Salvage Company, about three months later, it is short of funds. The *Blakely* never left Philadelphia because "the investing public has not purchased enough stock in the enterprise."

It was reported that "the treasure, if obtained, would be subject to a multiplicity of lawsuits which would keep it in the maritime courts and before the Hague and State Departments for half a century. Leavitt, however, waved aside the legal and engineering difficulties as trifling." Perhaps the potential investors appreciated the legal obstacles better than Leavitt did.

International politics muddied waters that were already obscured by judicial uncertainty. The Already Salvage and Towage Syndicate asked the Admiralty to protect its salvage rights, while the Lusitania Salvage Company asked the U.S. State Department "to safeguard its interests in seeking the sunken treasure on the ground that it was more than three miles off shore and therefore public property."

While this wrangle was playing itself out, Germany jumped into the fracas with its own absurd accusations. Germany wanted to have "expert German representatives present throughout the raising of the *Lusitania* and recovery of its cargo." Why? Because Germany alleged that the true purpose of the salvage operation was to remove all evidence of contraband, and thus rob Germany of its justification for torpedoing the *Lusitania* as a legitimate target of war.

Just as a few thousand pounds worth of valuables was inflated to $6 million in gold and jewels, German claims of contraband increased from rifle cartridges and fuses to "a cargo of munitions, torpedoes and two submarines." It is hardly possible that a couple of submarines could have been loaded aboard the liner in New York without some inquisitive bystanders noticing them.

In August, Lusitania Salvage Company office manager W.J. Sheehan told reporters of a change in plan. "It is so late in the season now that by the time we could get at the *Lusitania* the weather might be too rough to allow the diving to proceed, so it is practically decided to go on to the Mediterranean. It is smooth enough there to allow work to go on all winter.

"The best ship sunk in the Mediterranean is the Japanese ship *Yasaka Maru*, which went down with $12,500,000 in gold on board. Then there is the *Geelong*, which was sunk with $6,000,000 worth of jewels belonging to the Maharajah of some country. They are both in shallower water than the *Lusitania*." The phrase "of some country" demonstrates either unconscionably poor research or intentional disinformation.

According to records in Lloyd's Register of Shipping, the *Geelong* carried lead and general cargo at the time of her loss. No cargo is listed for the *Yasaka Maru*.

Asked about the company's financial situation, Sheehan replied, "There is still $30,000 worth of stock that has to be sold before we are ready to start. We have sold $95,000 worth of stock already, but we have to dispose of the rest first."

Leavitt was not only the inventor of the rigid diving suit, he was "the chief promoter of the stock-selling campaign." Robert Davis, managing director of the commercial diving outfit Siebe, Gorman & Company, inventor of the Davis Observation Chamber, and author of *Deep Diving and Submarine Operations*, wrote of Leavitt's diving armor, "It's utility under heavy pressure is questionable."

Also questionable was whether the Lusitania Salvage Company could continue to float its loans and salvage its own debts. When the company could not afford to pay the captain of the *Blakely*, and he refused to accept company stock in place of cash, the salvage vessel was left without a master.

Ultimately, the company's proposed scheme faded into obscurity.

PERMANENT INFLATION

One direct result of Leavitt's promotional activities was to fix in the collective mind of the press - and therefore of the public - the value of the *Lusitania's* treasure. By July 1923, hearsay phrases were bred out of the specie. Witness this front page announcement printed in the *New York Times*: "The *Lusitania* on her ill-fated voyage carried $5,000,000 in gold in the strong-arm room, and in addition the purser's safe was said to contain approximately $1,000,000 in gold and jewelry." Only the worth of the purser's safe was in doubt.

This blurb accompanied the news that Count Landi, one of the owners of an unnamed salvage outfit, left Dover aboard the steamship *Semper Paratus* "with clearance papers for the North Atlantic, reputedly to attempt salvage of the gold from the sunken *Lusitania*."

No follow-up stories appeared.

RAISING THE HULL

Edward Cassidy had a more grandiose scheme than recovering only the gold and jewels. He proposed to salvage the entire ship. He claimed to have developed "an entirely new application of an old principle for raising sunken vessels intact. A working model of the apparatus was made to an exact scale, predicated on raising ships of all sizes up to a tonnage 25 per cent. greater than the *Lusitania*. This model apparatus was tested thoroughly on a scale model ship of the *Lusitania*, at a comparative depth of 500 feet. In all, 150 tests were made, the sunken model being placed in three different positions to make the test as difficult as possible, and of the 150 tests, the sunken model was brought to the surface intact 148 times. The tests were carried on in a muddy pool, so that the work was done blindly, the sunken model being located, marked and raised under methods which would have to be employed in case of a vessel resting in water too deep for ordinary diving apparatus. Two weeks from the time the apparatus reached the spot would be quite ample for bringing the *Lusitania* or any other vessel of its size to the surface without damage. With this apparatus, the extent of injury to the hull of the vessel and the position in which it may lay are immaterial.

"Based on these tests, made under conditions fully as difficult in proportion as would be experienced in actual work, it is the opinion of the writer that sooner or later the *Lusitania* (if not destroyed in the meantime in attempts to secure its cargo) and other vessels with valuable contents will be brought to the surface intact. While the hulls may not be worth anything except for scrap, yet the simplest way to bring the cargo to the surface will be to bring the entire vessel to the surface intact."

Notice that Cassidy did not actually describe his "apparatus." He merely made claims as to its efficacy. Nothing more was reported on his proposal to raise the *Lusitania* intact. Perhaps he found that raising a scale model from the bottom of a muddy pool was not the same as raising a steel hull weighing thousands of tons from the bottom of the ocean.

The Recognition of Exaggerated Claims

James Young, in a feature article published in the *New York Times* in 1924, took a more reasoned and reasonable approach to shipwreck salvage propositions than did those who invested money in the many undertakings then proliferating. He also fell prey to repeating declarations which were by then universal beliefs.

"One of the prizes of the war worth $5,000,000 to the man who can board her is the ghostly *Lusitania*, lying in 285 feet of water. The pressure at that depth will turn heads and make men giddy. Many plans have been proposed to raise the *Lusitania*, or at least reclaim her treasure. One of the great questions involved is the German charge that she carried ammunition. When it seemed, not long ago, that pontoons might supply a means of recovering her, the Germans were ready to ask for representation at the raising, to prove their allegation.

"The war was still under way when the first efforts were put forth to take back some of the wealth which Germany had sent to the bottom. Since 1914 it is estimated that the waters around Britain have yielded £50,000,000 in treasure. Not less than fifty companies have been organized in the British Isles to seek out treasure ships. Some of these have not advanced beyond the stage of an elegantly lithographed stock certificate, but ten or twelve of them are operating or ready to begin. What has been done with the *Laurentic* is inspiration enough to turn attention upon every wreck that strews the near-by waters."

Some £7 million of bullion was salvaged from the *Laurentic*, but the wreck lay at a depth of only 90 feet, where divers were able to work effectively and blast the collapsed hull apart in order to gain access to the bullion room.

Young noted wryly, and with unintentional prescience about the future of underwater treasure recovery operations, "Each time a salvage enterprise is launched a great howdy-do ensues about the ownership of treasure hulks. Shippers of the gold, underwriters, the Government, all take a hand. The English courts are full of litigation arising from these claims. The man who recovers a treasure cannot be sure that it is his own, unless he keeps a tight mouth and a closed pocket."

This observation is even more true today. Many unscrupulous people found that it was more financially rewarding to sue a successful salvor after he had recovered sunken riches than to invest in chancy rackets or to work honestly toward making recoveries.

Simon Lake's Tube

In 1931, famed submarine inventor Simon Lake and Captain H.H. Railey joined forces to form the Lake-Railey Lusitania Expedition. Lake told reporters that the wreck lay upright on the bottom at a depth of 240 feet, and that the top deck lay 175 feet beneath the surface - within easy reach of his "tube."

Lake explained how, from a surface support vessel, "we will send below a steel tube which will rest on the A deck of the *Lusitania*. The hydrostatic pressure there will be seventy-five pounds per square inch, but no one will feel it more than they do in a submarine. All must go down the tube through a long stairway, until they reach a room twelve feet wide, eight feet long, and eight feet high. The tube will be perfectly supplied with air from the pumps above. There will be no pressure for the individual.

"It is only necessary to walk down the flight of stairs, inside the tube, under normal atmospheric pressure, until one enters the observation or divers' operating chamber at the lower end. This operating chamber is fitted with look-out windows.

"Various motor-operated winches will be on the exterior, also two doors providing exit for the divers when it is necessary for them to go outside to do their work. Before these doors can be opened a pressure of air must be admitted into the chamber equal to the water pressure outside. . . . You might call the arrangement for preventing exit from this room foolproof, as it is impossible to open the divers' door until the sea pressure is equalized by admitting air to the compartment. When this pressure is equalized the exit door is opened downward and no water will enter the room. Instead, the air flows outward from the compartment.

"Once in the observation chamber, spectators will watch the work of those on the *Lusitania*. Through glass windows, and with the aid of powerful searchlights, all will know what is going on, except when the divers are outside the ship. Vision of from forty to sixty feet will be possible.

"Frank Crilley will be in charge of the diving operations, under my supervision. He has gone down as far as 308 feet, and he tells me he has no doubts about this plan of ours. It will be easy for him, he says.

"Once we get the end of the steel tube firmly planted on the top deck, Crilley will enter the air-tight chamber, and on leaving it will start to reconnoiter. He has said that he can comfortably work for an hour at a depth of 200 feet at any time, and the fact that he can start freshly from the tube, with an hour for work, will be of immeasurable help. Usually most of the time a deep-sea diver is taken in being lowered and brought up, and this will be eliminated.

"The submersible tube for the diver's operating compartment is shut off from the rest of the apparatus by means of an airlock, which permits ready passage to and from the surface vessel under normal air pressure; it is only in the diver's compartment, where the air is under pressure equal to the compartment's depth of submergence, that the exit door is opened. This arrangement eliminates the 'bends,' ordinarily a great danger when men are under water.

"After Crilley has put in a shift, he need only come back to the compartment for a rest.

"We know exactly where the safe is and we will bring it out from the side of the ship by a method which I cannot explain at this time, secure it carefully with heavy chains and hoist it by derrick to the mother ship.

"From then on, prevailing circumstances will govern our activities. We wish to get everything of value, actual or sentimental, and bring it back to the world above from the sunken *Lusitania*. And considering that we shall be able to utilize about ten days in any winter month, we expect remarkable photographic results. The water in winter is more free from certain kinds of marine fauna, such as jellyfish, than in summer."

Simon Lake's tube was no imaginary invention. Lake explained, "Twenty years go I invented a tube designed to recover the treasure of the *Lutine*, a British frigate sunk off Holland's Zuyder Zee with $6,000,000 in gold bullion aboard. Lloyd's of London has kept an eye on the treasure for 150 years. The war stopped that enterprise, for I was called back to build more submarines for the United States Government. I recently started to operate under contract, building the apparatus for the *Lusitania* venture. The under-water part of it was thoroughly developed and has been proved in recovering cargoes from a number of ships sunk at a lesser depth than that of the *Lusitania*."

Although Lake never got back to the *Lutine*, in subsequent years he utilized the tube to recover coal and cargoes from wrecks off the Connecticut coast. The depth of the *Lusitania* required the addition of a swivel joint, which connected the upper length of tube with an extension tube that reached down to the level of the deck.

It is interesting to note that one columnist, in editorializing Lake's bold advocations, noted parenthetically, "It is doubtful if the *Lusitania's* strong-room and hold contain enough of intrinsic value to justify the great expense that Mr. Lake and his associates seem willing to incur." The *Lusitania's* actual value was known. Only those who chose to accept unsubstantiated hype as truth, or deluded themselves by believing in chimerical fortunes, fell prey to the lure of phantom treasure.

Lake and Railey were granted official sanction by the British Board of Trade, the Admiralty, the Foreign Office, and the Liverpool and London War Risks Insurance Association. None of these agencies showed any concern that Lake and Railey would discover anything on the wreck of a culpable nature, such as guns or munitions. The only caveat given to the expedition was that the company should report recovered items "to the Receiver of Wreck at the nearest custom house and give proof that the salvage was effected by arrangement with the association."

Lake's and Railey' goals were realistic. Railey: "No attempt will be made to raise the *Lusitania* and we are not equipped to undertake general salvaging of the vessel's cargo, which, intrinsically, would barely justify the cost. We propose purely as a demonstration of the revolutionary aspects of the Lake submarine salvaging tube merely to bring to the surface the ship's safes and other miscellaneous articles of actual or historic worth."

Railey expressed confidence that "the expedition will succeed in taking motion pictures and still photographs of the condition of this famous wreck after sixteen years' submersion. The important research in submarine lighting will be conducted in cooperation with the Westinghouse Lamp Company of New Jersey." Special 5,000-watt lamps developed by Westinghouse had already been tested under water; they were used in a trial run to illuminate 150 feet of film shot in shallow water off the New Jersey coast.

Despite the amount of copy that Lake and Railey received, and official approval, the expedition was never mounted.

The Sorima Company

Deep-water salvage hit the headlines when the Italian salvage outfit Sorima set its sights on the *Egypt*. The *Egypt* was sunk in a collision in 1922 off the coast of France, and settled to a depth of 400 feet. Because of the large quantity of gold and silver aboard (worth about $5 million), a long search and salvage operation was conducted and eventually proved successful. The search took two years, salvage another five. The job was completed in 1935.

No divers were employed. Instead, Robert Davis designed a pressure-proof observation chamber that was lowered to the wreck from the *Artiglio*. Through telephone lines an observer inside the chamber directed the placement of explosives and the maneuvering of a grab or claw which scooped exposed bullion out of the wreckage.

Minus the lure of fantastic wealth and a worthwhile reward for their investment, commercial salvors spurned the *Lusitania* because they knew the truth about her cargo of gold and jewels: there wasn't any. The war left the ocean floor littered with shipwrecks whose cargoes were more valuable and more easily obtainable.

The *Lusitania*'s grave was yet unmarked. In 1934, it was reported that the Sorima company placed buoys "at a wreck close to where it is believed the liner foundered . . . about fifteen miles south of Old Head of Kinsale. The *Artiglio* and her sister ship, the *Arpione*, after placing the buoys, continued sweeping to locate a number of vessels that sank while carrying copper. Diving will begin in calmer weather."

Sorima company officials denied that their salvage vessels were "in Irish waters to attempt to salvage the *Lusitania's* gold." Perhaps they intended instead to salvage the liner *Minnehaha*, which sank only a few miles from where the *Lusitania* went down. Nothing more was reported about Sorima's Irish operations.

Diving Armor, Jim Jarratt, and John D. Craig

In his exciting adventure book *Danger is My Business*, published in 1938, John D. Craig related information that was given to him by Joseph Peress, inventor of an armored diving dress. According to Craig, "the Tritonia Corporation of Scotland directed by H.J. Demetriades, a Glasgow capitalist originally from Greece, took a hand. This corporation obtained from the Liverpool and London War Risks Association, Ltd., the British government insurance agency, a contract to salvage four hundred ships lost in the World War. The *Lusitania* was one of them."

Using the former lighthouse-service boat *Orphir*, Tritonia spent nearly four months during the summer and fall of 1935 locating shipwrecks by means of a depth recorder. The culmination of numerous wreck discoveries was one which was graphed at 769 feet in length and which towered 84 feet above the sea bed.

The depth to the bottom was 312 feet. Jim Jarratt was lowered to the wreck inside a Peress one-atmosphere suit that was like the Davis observation chamber except that it possessed articulated arms and legs. The steel arms terminated in claws which could be manipulated by hand controls.

At 240 feet "the winches stopped creaking, and Jarratt spoke over the telephone: 'I am standing on the plates of a ship, but under it there is little corrosion. I will measure the rivets.' He did, then telephoned. 'They are about two inches.' The rivets of the *Lusitania* were one and seven-eighths inches in diameter."

Then, due to a three-knot current, the *Orphir's* anchor pulled out of the wreck and nearly smashed into Jarratt where he dangled helplessly on the hull. He was pulled up quickly. Jarratt did not have to decompress because he was not under pressure. "He never got down again. Fog and rough seas closed in for the winter. The *Orphir's* task was finished for the time being. She had located the *Lusitania*."

Proposed Filming Expedition

Craig reiterated what was then most commonly and mistakenly believed about the *Lusitania* after the truth of the event had been muddled with contrary claims and passed down through a generation of retelling: "The Germans said she had been torpedoed because she carried munitions. The British government denied it. The world wanted to know. And there was reported to be from two to fifteen million dollars in gold bullion in her strong room, plus three hundred thousand dollars in currency and jewels in the purser's safe."

As stated in the previous chapter, Germany knew nothing about munitions aboard the *Lusitania* until England volunteered the information. There was no secret about it. Nor was any mention ever made of a cargo of bullion, either before or after the sinking or in the voluminous investigations and follow-up insurance claims. Quite the contrary, if bullion were being transported across the Atlantic Ocean it would have been headed west, not east, in order to pay the United States for much needed food, fuel, and general supplies.

By 1938, most Americans believed that the sinking of the *Lusitania* was what prompted the U.S. to declare war on Germany. The two-year hiatus between sinking and declaration had been forgotten by the general populace.

In any event, Tritonia hired Craig to photograph the *Lusitania*. Craig had a great deal of experience in organizing expeditions to photograph wildlife in Africa and marine life in diverse underwater environments. He contracted the General Electric Company to build eighteen lamps of 5,000 watts each. It took the company a month to design, construct, and test the watertight housings, bulbs, and electrical connections. In tests, the units withstood a pressure equivalent to that found at a depth of 1,500 feet.

Craig decided that the Peress armored suit was too cumbersome for his needs. He and his long-time companion Gene Nohl designed a hard-hat outfit with a radial face plate for all-around vision, a padded breastplate, a heavy-duty canvas dress, and chafing pants with pockets for tools. The helmet was fitted with a hook so it could be lowered over the diver's head and so the diver could then be lowered into the water like a piece of bait on the end of a fishing line.

"In the helmet we also built a dashboard, with a compass, depth gauges, watch, microphone, receiver, transmitter, and other instruments. We even included a container for food, in case long dives were made. . . . On top of the helmet we built an air purifier canister, protected against wetting. On the dashboard we mounted an auxiliary air purifier, just in case. The regular canister is good for twenty-four hours of diving, the auxiliary for one hour."

In order to offset the narcotic effect of nitrogen under pressure, Craig decided to replace the nitrogen with helium. He put scientists to work devising decompression schedules for mixtures of oxygen and helium, and working on the problem of oxygen poisoning if he had to breathe air in case the scientists could not solve the helium problem in time.

"Now the *Lusitania* diving was called off for the summer of 1936. Hitler had marched into the Rhine. Mussolini was invading Ethiopia. The British government considered it best to let sleeping dogs lie. To salvage the *Lusitania*, with its attendant publicity, would be rubbing salt into German wounds. They wouldn't like it. So the *Orphir* went to reconnoiter some of the ground where other vessels lay, and we were advised to continue our experiments."

Craig and Nohl continued their design work and experimentation. By 1937 the unrest in Europe was growing by leaps and bounds. The expedition was once again postponed. Craig and Nohl conducted tests in pressurized chambers with breathing mixtures of helium-oxygen - using themselves as guinea pigs. On December 1, wearing the Craig-Nohl dress and breathing "helium air," Nohl "established a new official world's record with a dive of 420 feet in the waters of Lake Michigan."

All these preparations came to naught. The global political situation was precarious, and war with Germany again loomed on the horizon. Hitler's demands and the Nazi threat made 1938 appear ominous. Then came 1939 and the German invasion of Poland. Tritonia's and Craig's plans to dive the *Lusitania* were permanently quashed.

ROYAL NAVY DIVES IN 1948: TEDDY TUCKER AND RISDON BEASLEY

Rumor and speculation still abound about clandestine diving operations that were conducted on the *Lusitania* by the Royal Navy in 1948, in a supposedly belated attempt to avoid any scandal that might result from the possible discovery that the liner had been armed. A great deal of copy has been generated throughout the years about this secret cover-up mission, whose objective (it was believed) was to blast open the hull in order to remove concealed naval ordnance.

It is difficult for me to accept that after two all-out wars against malevolent German governments, anyone in the Royal Navy or modern British government could possibly care about a technical indiscretion from more than thirty years before, even if it were true. Conspiracy proponents would have it otherwise. They want to believe that the Navy salvaged the guns and then absconded with the evidence; or, failing that, that the Navy depth-charged the wreck in order to destroy what could not be removed.

While it is true that the Royal Navy has been reticent about discussing its operations on the *Lusitania*, and did not publish an account of its underwater maneuvers, it is also true that such security is nothing more than normal. Military operations are covert by nature. Yet people have chosen to let standard operating procedure feed their imaginations and inherent desires to believe in government conspiracy, in the same manner in which current ufologists accuse the U.S. Air Force of covering up incidents of flying saucer landings and alien abduction.

Also circulating in the rumor mill is the charge that famous Bermudan treasure salvor Teddy Tucker dived the *Lusitania* when he was in the Royal Navy. Tucker repudiates that claim. In a recent interview he told me that he did not know how such a story got started, but hoped that I would help to put an end to it.

In truth, the Royal Navy did conduct some dives on the *Lusitania*. Tucker was a diver in the Royal Navy at the time, and he was a member of the team that carried out the diving. But he insists that he did not personally descend to the wreck; he acted solely as surface support. Nor was the operation part of a super secret mission. Half a dozen British hard-hat divers were put down on the wreck with instructions to look for guns that might be protruding from the hull - not so they could be secretly salvaged, but instead to satisfy historical curiosity and to put the trumped-up rumor to rest. They did not find any guns, nor signs that any guns had been surreptitiously removed.

Through time and due to neurotic fixation, a minor Naval exercise of short duration has grown to large-scale mythical proportions in order to bolster the claims of conspiracy fanatics. I have no doubt that even if the notes and records of the Royal Naval expedition were published, they would not be believed by those who have already reached their own contradictory conclusions. The heart of conspiracy theory is that claims are by nature unprovable.

It has also been alleged that commercial salvor Risdon Beasley salvaged guns from the wreck under the auspices of the Admiralty at a time coinciding with the diving operations conducted by the Royal Navy. Again, this is mostly speculation. Beasley has denied diving and salvaging the *Lusitania*, although it is possible that he helped the Royal Navy to relocate the wreck.

In conclusion, any diving that was done around this time accomplished nothing. Nor does current observation of the hull bear witness to telltale signs of commercial salvage beyond that conducted in 1982, in the vicinity of the specie room.

Depth-charging the Wreck

The assertion has also been made that the Royal Navy depth-charged the *Lusitania* in an attempt to destroy any evidence of armament. Again, it is patently absurd that the British government would go to such lengths after the Second World War, to cover up an issue that was long since dead - especially in light of Germany's two-time culpability in the commission of human atrocity. There can be no doubt in anyone's mind that the abominations of the Kaiser and his military forces can ever be justified by so trivial a discovery as a gun stowed deep in the *Lusitania's* hold.

This is not to say that the *Lusitania* was not accidentally depth-charged during World War Two, when deadly U-boats patrolled off the British Isles and when the sunken liner may have looked like an enemy target on the primitive sonar sets employed by antisubmarine warfare units. Nor is it not impossible that the *Lusitania* - as well as other wrecks in the vicinity - was used as an easily locatable target on which to test improved sonar systems and depth-charge accuracy.

Furthermore, although the pressure hull of a submarine can be ruptured by the nearby explosion of a depth charge (most effectively when detonated underneath the hull, since most of the explosive force is directed upward), depth charges are nearly useless for clearing underwater obstructions such as ships' hulls and rock piles. For that purpose, charges must be precisely placed. Demolition experts in the Royal Navy undoubtedly were aware of this.

As noted at the end of the previous section, current observation of the wreck shows no sign the use of explosives except in the vicinity of the specie room in the stern.

Risdon Beasley

Rumor still runs rife in Kinsale that in the early 1950's, commercial salvor Risdon Beasley lowered charges down to the wreck and commenced blowing it apart. Beasley has been even more reticent to discuss his activities than the Royal Navy, going as far as to deny all "charges." For decades, his outfit salvaged shipwrecks off the Irish coast. He may have begun salvaging the *Lusitania*, or he may simply have been working other wrecks in the area. Commercial salvors seldom advertise their exploits unless they are trying to attract investors.

Publication Revival

1956, *The Last Voyage of the Lusitania*, by A.A. Hoehling and Mary Hoehling. Nearly forty years had passed since the last book was written exclusively about the *Lusitania*. As I have already noted, until the 1950's, shipwrecks did not possess the mystique or attract public attention as they do today. They were catastrophes that were held no different from natural disasters such as fires, earthquakes, and volcanic eruptions.

The 1953 film *Titanic* and the sinking of the *Andrea Doria* in 1956 undoubtedly contributed greatly to the popularity of such disasters because they were seen by millions in the movies, on television, and in newspapers and magazines.

Following the film *Titanic* came Walter Lord's book, *A Night to Remember*, which covered the same material but with attention to truth and detail. The film was a fictionalized portrayal, whereas the book was an accurate rendering of a dramatic episode in the lives of the rich and famous. This was perhaps the beginning of the perception of shipwrecks as discrete historical events.

The Hoehlings' *Lusitania* book is a factual account that follows to some extent the mold of Lord's *Titanic* book, although, to be fair, not much variation on the theme is possible. I would not go as far as to declare that shipwreck tales are formulaic, but certainly the circumstances of a sinking ship seldom leave much room for creative telling. A description of the sinking, the death of the unfortunate, and the struggles of the survivors are what all shipwreck stories are about.

The Last Voyage of the Lusitania does not delve into histrionics or make unsubstantiated claims of conspiracy. The book is a well written, balanced account whose popularity has not dimmed throughout the years. It has been reprinted many times since its initial publication, including mass market paperback editions.

At the very least, the book introduced the *Lusitania* to a new generation, and kept the name alive in the public mind.

John Light - A Man Ahead of his Time

Take the gruffness of Humphrey Bogart's many movie characters, combine them with audacity, arrogance, and self-assurance, add a pinch of devil-may-care attitude, and soften the brew with familial love and obligation, and you'd have a rough portrait of John Light's personality if not an insight into his soul. Light could be brusque and short-tempered, especially with people who tried to delve into his affairs. He was a no-nonsense fellow who did not suffer fools. He was also solitary, energetic, and enigmatic - a queer admixture that kept people guessing what he was truly about. He was not a person who was easy to get to know, nor someone you might want for a friend.

Yet there was something about his character that set him apart from everyday people. Perhaps it was his drive, perhaps it was a strident desire to prove a point - or perhaps it was a need to confirm his identity. Light was not a talker; he was so curt and taciturn that he gave new meaning to the words. He also put his life and livelihood on the line for a goal that eventually became a yearning and lifetime ambition - to wrest from the *Lusitania* her deepest secrets and treasure.

"There is no gold," Light stated bluntly and continuously in interviews. "How often do I have to tell newspapermen that that's bunkum?"

To Light, the *Lusitania's* treasure was her history. If he had any faults in this perspective, it was his ingenuous belief that the liner's untimely end was shrouded in diabolical intrigue and chicanery. His ardent desire was to prove that the *Lusitania* was armed at the time of her loss. "A lot of pretty important people had insisted she carried no guns: the President of the U.S., the Secretary of State, the Collector of the Port of New York, the Cunard bosses and the Admiralty brass. Gustav Stahl went to Atlanta penitentiary for saying she did. Even at this late date it might be embarrassing if evidence turned up to show that these important people were wrong - or were lying." (Stahl was a German national living in the United States. He testified that he saw guns being mounted on the *Lusitania*, and was convicted of perjury.)

I dare say that all the prominent people who were mentioned by Light were dead, and in that condition would prove difficult to embarrass. But the pronouncement made good newspaper copy and might even attract investors in the form of movie or television producers. Nor can I rule out the possibility that perhaps Light desperately needed to be right at someone else's expense: a better-than-thou sentiment that bespoke some hidden insecurity.

John Light started his diving career in 1949, when he was only 16. He joined the U.S. Navy after graduating high school. During his two years on active duty he became a submariner and helmet diver. After

66 - The *Lusitania* Controversies

mustering out of the service he worked as a civilian salvage diver. In 1956 he became a cameraman, in which capacity he worked freelance for the National Broadcasting Company. By 1960 he had gained the proficiencies necessary to tackle the goals he had set for himself to accomplish: to film the *Lusitania's* grave site and to uncover solid evidence of erstwhile British improprieties.

Light first dived the *Lusitania* on July 20, 1960. He wore a wetsuit, mask and fins, scuba cylinders, and a double-hose regulator. He breathed ordinary air. Alone he descended to the upper hull, which he reached at a depth of 245 feet. At the age of 27, he was only the second person to touch the liner's encrusted steel plates - twenty-five years after Jim Jarratt descended to the wreck in a one-atmosphere suit.

Several weeks after his first solo dive, Light returned to the wreck with a couple of stalwart deep-diving cohorts he had met during filming assignments, and began a methodical exploration of the outer hull and superstructure. He then convinced others to join the venture. He dived the wreck sporadically for the next two years, at the end of which time more than one hundred dives had been conducted, of which twenty-five were made by Palmer Williams and thirty-eight were made by Light himself. According to Dennis Morse, another member of the diving team was Chuck Osborne.

Wrote correspondent Kenneth MacLeish: "The exploration of the rusting ruin had changed for Light from a normally exciting deepwater film project to a personal crusade, almost an obsession."

What primed Light's obsession was "a long, tapered object" that "could have been a gun, or a spar, or a pipe." Light's partner on that dive, U.S. Naval officer Surgeon-Commander John Aquadro, said, "I saw what I felt was a gun barrel and questionably a gun emplacement or turret near by: a semispherial piece of metal, riveted. I saw the gun just as it was time to come up, but I did take a quick second look."

Light was not as close to the object as Aquadro, but he confirmed that it lay at the bottom of "a hole some eight feet across, black and bottomless. . . . as if some heavy object had dropped straight down after the ship had settled on her side. Maybe long after. The superstructure is made of light materials, and it looked as if something had just torn on through. And there were three steel cables leading into the hole that had been drawn so tight that they had sliced right into the surrounding metal. Maybe they were attached to whatever had fallen through. Maybe somebody was trying to hoist that thing away, and it broke loose."

The latter statements refer to his belief in the Royal Navy's attempted gun recovery operation.

MacLeish described his impressions of the descent he made with Light and Williams in 1962: "The anchor cable rasps through one gloved hand as we hurried on our vertical and inverted way. It gives a measure of motion; without it we could not sense our rate of descent: 100 feet, twilight and deep chill; 150 feet, a turbid zone where light and temperature drop sharply; 200, clear again and bitter cold, with the faint sourceless illumination of starlit midnight; 250, the air is thick in the throat and flows like a light icy liquid into the lungs. Lips tingle with narcosis, and thoughts blur. The eyes need a point of focus to steady the senses, but there is no fixed form anywhere. Watch and depth gauge help. Bright and solid, they are points of reality in this imprecise void. With a little effort both of them can be read: one and a half minutes since starting down; 260 feet."

I find it difficult if not impossible to believe that they made their descent in ninety seconds. MacLeish's imagery of narcosis was conveyed more bluntly by Palmer Williams, who told Howard Klein that he was "narked out of his mind." *That* I can certainly believe. MacLeish wrote that one diver "was so dazzled that he surfaced with no memory of the dive he had made - and never made another."

MacLeish was an experienced deep diver, a careful observer, and a vivid writer who could transcribe events with deep emotional appeal: "All at once there is a lightening below. . . . Where are we? Deep between walls that tower high above us. But, of course, they are not walls but decks, for the ship lies on her starboard side. We are inside her smashed superstructure. There are remnants of wood flooring attached to a vertical surface near by. There is a warped railing. There are other shapes, disoriented details of an unseen entirety. . . . Four minutes gone, 275 feet: we drop toward tangled wreckage below, down another 15 feet, but there is nothing intelligible there. Best to soar up out of the chasm and see what lies above. There the broad expanse of the hull stretches away, smooth and undamaged save for a row of half-sheared rivets. Portholes pass in orderly sequence." Elsewhere MacLeish (and Light as well) emphasized that the wreck lay on its starboard side and that the decks were vertical - an observation that is at variance with the true angle of list, which is about a 45°.

Light also claimed that the bow of the *Lusitania* was nearly severed from the main hull, and that this catastrophic separation constituted proof of a secondary explosion that was the result of explosives stowed in the locality where the torpedo struck. This observation was not borne out by subsequent expeditions (in 1982, 1993, and 1994), nor is it currently accepted. Was John Light narked, or suffering from wishful thinking?

Unfortunately, MacLeish was severely bent after the dive he just described. He was partially blinded, lost his sense of balance and motor control, and experienced other neurological symptoms. He spent twenty hours in an emergency recompression chamber.

Although his life was saved, his recovery was only partial. He underwent further treatment in a hospital hyperbaric facility.

Said Light afterward, "Season's finished. So are we, for now. Right where we were last year. The payoff may be only two good dives away. I think the Navy officer will swear that he saw the gun. But we have to get a picture first. then we can draw up the affidavits. That means we'll have to come back next year. We're too close to the end to quit. We're just too close."

It was as close as Light ever got. Never again did he dive the *Lusitania*. He kept coming back to Kinsale, and he lived there for a while. He even married a local lass, Murial Acton, and fathered four children with her. In 1967, he went so far as to purchase the salvage rights. And he maintained his obsession for the wreck until his death thirty years later. As the dry decades slipped away, he continued to delve into archival research, and he dreamt optimistically of the day he would return and make his grand hoped-for discovery of British bureaucratic perfidy.

While John Light's dives were indisputably a triumphant tactical success, they were also a sad strategic failure in that he failed to achieve his goals or to add any new knowledge about the wreck. This statement is not intended to belittle his unremitting effort nor to denigrate his accomplishments. After all, he conducted his dives on scuba at a time when the equipment available was primitive, when training was inadequate, and when so little was understood about the characteristics of shipwrecks and the special conditions they create.

The weather in the area can be abominable. At times, Light was lucky to dive on one day out of five. Many dives were spent in securing descent lines to the wreck. These lines had a knack for chafing through and going adrift, or parting during a tempest, or being stolen by local fishermen.

Dives could be made only during slack tide. This occurs four times daily and lasts for about three-quarters of an hour, but only two slack tides occur during daylight. When the tide is running, the water moves so swiftly that one cannot swim against it or hold onto anything against its onslaught.

Bottom times were limited to ten minutes, minus the time spent descending to the wreck. If MacLeish is to be believed, and they reached the upper hull in a minute and a half, then the divers had eight and a half minutes to explore the wreck, shoot film, take pictures, and make observations - all while severely handicapped by nitrogen narcosis. In reality, and under normal circumstances, they probably took three to five minutes to descend to the upper hull. Thus the effective working time of each dive lay in the range of five or six minutes. This did not permit much time for exploration, and it is doubtful if they ever traveled more than a few feet away from where they touched down.

Decompression was conducted in the water, mostly while hanging onto the descent line. Total decompression time was about half an hour, beginning at forty feet. The final ten-foot stop was conducted inside a bell that was suspended in the water. The diver was wet from the chest down, but he could remove his regulator and breath oxygen through a mask.

The camera housings Light used were heavy and bulky and awkward to carry. Natural light seldom bathes the wreck, so in the near absolute blackness he had to rely on artificial sources to illuminate his subject. Existing lighting systems were dim and not self-contained: lamps had to be powered by topside batteries, which meant that hundreds of feet of waterproof cable had to be lowered down to the wreck. These cables often leaked, which meant that the lamps did not work and the dive was wasted. Even when the lamps functioned as intended, the deadweight of the thick cable tethered the lighting system to the descent line like a dog on a leash, making it impractical to explore more than a few feet away.

After two years of work - from 1960 to 1962 - Light shot no usable movie footage and took no publishable photographs. It is true that he was ahead of his time - but too far ahead. He was not able to overcome the conditions found on the wreck, or the limitations of dive gear and camera equipment that was available to him. Furthermore, his outlook was blinded by his single-minded conviction that the *Lusitania's* loss was due to a fiendish political stratagem that was intended to embroil the United States in the war against Axis powers.

With all due respect, I must also add to the list of John Light's deficiencies the fact that he was not a wreck-diver. By this I mean to impart that he did not understand shipwrecks: how they collapse and deteriorate, how their presence alters the environment, what special techniques must be utilized to explore and understand them. Light cannot be faulted for his lack of understanding: wreck-diving as a specialty did not exist at the time.

Which presents the perfect segue into the next Part.

Part 3
1950's: The Origin of Wreck-Diving

THE BEGINNING OF SCUBA

Perhaps the first avid wreck-diver to fit the category was Michael de Camp, a grade school science teacher from Morristown, New Jersey. De Camp was introduced to scuba by his friend Carleton Ray. They met initially at Berkeley, California when they both lived there, and where Ray was working on his doctorate in marine zoology. After earning his Ph.D., Ray accepted the job of assistant curator at the New York Aquarium. De Camp soon moved back east, so their friendship was easily maintained. Later, when Ray went to Nassau with Elgin Ciampi to write an underwater guide to Caribbean coral reefs and the fish that inhabit them, he invited de Camp to come along.

The year was 1954. At that time certification courses did not exist. Nor were there dive charter boats, specialty dive shops, or air filling stations. The only formalized instruction (note the singular) that de Camp received from Ray was "don't hold your breath."

De Camp already owned a mask, a pair of flippers, and a snorkel. To be a fully equipped scuba diver he needed to purchase only two more items: a steel tank and a double-hose regulator. He paid a visit to Richards Army and Navy Store on 42nd Street in New York City, and emerged with the newly imported, newfangled equipment. Wetsuits had not yet made their appearance on the market, so when he reached the Bahamian shore that summer and approached the ocean spray, he was clad only in swimming trunks and a long-sleeved shirt. He lashed the tank to his back by means of nylon straps which were secured to a canvas bag which fitted tight around the tank like a canteen cover. He placed the regulator mouthpiece between his lips and slipped silently beneath the waves. Slowly he inhaled and exhaled; he remembered Ray's advice not to hold his breath.

De Camp was a bona fide scuba diver. He was hooked on diving as if it were a drug.

In succeeding years, de Camp gratified his newfound habit by diving in local lakes and quarries and in the nearby coastal shallows, gaining experience under water and confidence in his gear. In those early days of scuba, getting tanks filled with air was a problem. Only a few commercial establishments had compressors that could pump air into cylinders that were then considered high-pressure: 2,250 pounds per square inch. Fewer still had filler assemblies adapted to

1950's: The Origin of Wreck-Diving - 71

the valves of scuba tanks. Hardware stores or pool accessory shops were the mainstay. But he persevered with the inconvenience of traveling long distances to obtain air fills in order to continue his addiction.

Wreck-Diving Genesis

Somewhere along the way he learned about the sunken wrecks that lay off New Jersey's sandy shores. Ships that lost their way in fog or encountered fierce storms off the turbulent coast often found their final anchorage within close distance to the beach. These shallow hulks from yesteryear were de Camp's first encounter with shipwrecks - and so began wreck-diving in the modern sense of the term.

De Camp was not the only one to discover the challenge and therapeutic benefits of recreational scuba diving: an outlet for urges which the regimentation of society often stifles. Other adventure-minded people turned their attention to the underwater world. As their numbers swelled, so did the demand for merchandise. Some sporting goods stores and hobby shops allocated a small bit of space on their display floors and shelves for the outlandish paraphernalia that was at first perceived as expensive and hard-to-sell playthings for a minority of the well-to-do in search of toys to collect - people who were thought of as eccentrics in pursuit of a fad. Against all prediction, an exponential growth in sales was spawned by manufacturers with new and different products to peddle, and was spurred by tourist resorts with leisure time activity to promote. Warehouses were stocked with inventory for willing buyers clamoring for the newest piece of equipment or the latest modification.

This self-fueling cycle continued until diving became an honored and accepted pastime more fulfilling than mere amusement. Retail stores that specialized in dive equipment and accessories, and that sold nothing else, began to appear: the dedicated dive shop was born. Full scale wreck-diving lay not far behind.

Birth of the Second Generation

While de Camp was honing his skills in offshore ocean diving and learning the art of underwater photography, this author was attending grade school in the Philadelphia suburbs. I was a baby boomer whose first scream was heard in the neighborhood where my father had been born and raised: South Philly. When I was about two years old, Domenic and Meta moved to my mother's hometown along the eastern shore of Maryland, in the quaint farming community of Salisbury. We stayed only a year, returned to South Philly, but soon emigrated to the northeast section of the city, where I still reside.

Both my parents worked full time: my father as an electrician and later as an electrical contractor, my mother as a private secretary. So I went to nursery school and kindergarten and, when I started grade school, stayed after classes until my father arrived after work to take me home.

Nowadays there is controversy over the emotional damage done to children who attend day care centers instead of spending more time with their parents. Since no one ever told me that I was being deprived, I never felt it. I always thought that day care school was great fun. It gave me time to do my homework, there were other kids to play with, we had crafts and art and toys, we played kickball outdoors in the warmer weather, and the teachers introduced us to many novel and creative amusements such as finger painting and molding models with self-hardening clay. My special enjoyment was making clay dinosaurs which were then shellacked for a fine dark finish. It was worthwhile time well spent.

What unfortunately comes to mind about my early school days is a regrettable episode in which, to my shame, I eventually participated. One little girl came from a family so poor that she attended school wearing a dress that was frayed and faded and in shoes with soles which were peeling off her feet; her dirty blonde hair hung straight and stringy. The boys took to taunting her and calling her "cootie." She never retaliated in any way. She accepted the repeated jeers silently while returning a staid, pitiable stare; perhaps it was a look of helplessness. Once I got swept up in the frenzy of the moment and joined the other boys in mocking her. Afterwards, I was aghast at what I had done. From then on, whenever I saw her in the hall I quickly looked away, embarrassed. I felt sorry for her and wished I could do something to make up for my reprehensible behavior. But I had neither the courage nor the maturity to apologize or to take up her side. I am still haunted by her sad, forlorn face. No one stuck up for her in her plight and she had no one to defend or befriend her. I can't erase that awful scene from my mind.

Suburban Philadelphia

Today people think of Philadelphia as another concrete jungle, but in the 1950's the Great Northeast, as the northeast section of the city came to be known, was so far out in the suburbs that there were more groves than lawns, more farms than yards, some of the local streets were yet unpaved, and many did not have sidewalks. I lived with my parents in an apartment till I was ten years old. Then, when my father's electrical business flourished, he built a new home and we moved into the single family dwelling in which I spent my formative years. The stone ranch house was bordered on one side by forest, on the back by an ancient railroad track across which lay a wooded marsh and a pasture occupied by grazing cattle, on the other side by a two story red brick house, and in front by cleared dirt which eventually became a road but which at the time was a flooded gully in the spring and autumn, dried and cracked mud in the summer, and a skating rink in the winter.

There were always squirrels and rabbits in the back yard, often pheasants and quail, sometimes white-tailed deer, rarely raccoons and opossums. Newts and salamanders lived in the creek that separated our back yard from the elevated railroad bed. Small fish, frogs, harmless snakes, and crayfish lived in the marsh and in the stream that meandered through it. Birds of all kinds landed on the cattails as if they were hitching posts. Today the marsh is a housing development and the forest is a shopping mall.

Growing Fears

I had two irrational fears as a small child: spiders and dark spaces. The idea of creepy, many-legged things crawling over my skin gave me shivers. Once when a granddaddy longlegs ran up under my pants legs I went into a fit of jumping and slapping to get it out. Notwithstanding the biological distinction that granddaddy longlegs are mites and not spiders, I was also afraid of centipedes and thousand-leggers, the operative characteristics of my fear being multiple walking appendages with a fast gait. I don't recall being afraid of millipedes, which had more legs but which were slow and predictable.

I did not like being afraid of things that reason indicated could not hurt me, so I decided upon a course of action designed to help me overcome my fears. I stood close to spider webs and watched spiders' activities (although I was always prepared to spring backward if the spider attempted to jump on me). After getting used to that I began touching spiders on the back with an admittedly shaky finger. I observed that they always scurried away from me instead of attacking. Then I spotted a granddaddy longlegs in the grass and deliberately put my arm down in front of it and let it scramble over. It gave me the shivers but I survived uninjured. Eventually, I could pick up spiders in my hand and watch nonchalantly as they wandered along my arm, transferred from one arm to another, or climbed over my neck and down my back.

The two primary figments of my fear of dark spaces were alligators that lived under my bed, and horrible monsters that lurked behind closet, attic, and basement doors. I slept with a night-light glowing softly throughout the night, and I was fanatically careful to lie in the precise middle of the bed, with my arms pinned to my sides and my legs straight. If I awoke and found my hand lying over the edge, I jerked it back in terror. If I had to get up in the middle of the night, I leaped from the comfort of the bed and landed as far away as possible; upon my return, I took a running start and jumped from the middle of the floor to regain safety. I could not get to sleep if the closet doors were not securely shut. Even in the daytime I stood at arm's length when I opened a closet door, fearful of what might spring out from the darkness within. I never turned my back on attic and basement doors; I pivoted in order to face them when passing.

In order to conquer these absurd fears I forced myself to stand close to closet doors when I opened them. I purposely put my back to attic and basement doors, fighting down the chills that ran rampantly along my spine. Eventually I was able to fling the wooden barriers wide and stare frightfully into the inky blackness beyond, although not without a ghastly dread of what was assuredly going to reach out and grab me. Finally, I walked down the basement stairs with the lights purposely left off. At first, I raced back up immediately and slammed the door behind me, my spine a racetrack for galloping goose bumps. As I grew bolder I was able to turn my back on the dark abyss for several seconds before running up the stairs. Then I went all the way to the bottom step and turned around, daring the creatures that dwelled in darkness to make their presence known. I was master of the situation when I could do this when the basement door was closed.

From Misfit to Maverick

I didn't fit in with most of the boys my age because I was physically smaller and less emotionally developed, a gap which widened with puberty. I could have been the stereotypic 97-pound weakling found in the Charles Atlas ads, except that I didn't weigh enough to qualify. I could best be described as "puny."

The neighborhood boys only rarely let me join in their social activities or organized outdoor sports. Tad Decker, from down the street, occasionally got me invited to play football when the boys were choosing teams and found an odd man out. I usually went along just to watch from the sidelines. The only position I could play was end; I was fast enough to outrun nearly everybody, I was so nimble-footed that I could dodge like a bat and stop on a dime, and I caught passes as if my hands were coated with flypaper. But when I got caught, I got creamed. One time I went down in a heap of bodies and accidentally caught a knee in the solar plexus. It knocked the wind out of me so bad that I didn't think I would begin breathing again till the middle of next week. The apologies were overwhelming, but I wisely decided that for one of my stature football was not the game to play. I became pretty much a loner.

The girls ignored me, but some of the boys picked on me because, with my stunted size, they could get away with it. Since I had a temper that lay somewhere between slow and nonexistent, riling me to anger was a difficult task to achieve. Nevertheless, as custom demanded, I returned name calling with equal intensity despite the greater mass of my antagonists. For some reason which I can no longer recall, one exchange of invectives escalated beyond the bounds of bluff, and I found myself committed to the proverbial school yard brawl. In those days boys did not gang up on a solitary opponent, but let two adversaries slug it out unaided. I didn't know anything about fighting, so the boys who rallied to see me win gave me a five minute crash course on how to make a fist so I wouldn't break my thumb, how to hold my arms in order to protect my body and face, and how to feint and lead into an attack.

Kenneth Stoker was a bully and the biggest kid in the school. He was a head taller than I and proportionately heavier, and his reach was so long that I never got in a decent jab. The makeshift ring bulged as we circled for position. For a couple of minutes Stoker and I sparred back and forth to little effect. We each had allies who cheered at every blow. Then he knocked me to the ground, jumped on top of me and straddled my body, and rained punches on my face until I was bruised and bleeding. His supporters had to pull him off before I was seriously hurt.

Solitary Solace

I loved to climb. When my mother wanted me to come in for dinner she would walk to the edge of the woods, look up, and yell for me. Invariably I was in the top of a tree, often in the one in which I had built a fort about thirty feet above the ground. One time a blackbird imitated her voice, and for years thereafter when that bird passed our way during its migratory travels, it called "Gary" during its temporary stay. Once I came home because I heard my mother calling incessantly, only to learn that supper was a long way off. The blackbird kept calling me anyway.

As I got older I traveled farther afield. I ranged deep into the woods, swung on vines across creeks, leaped like a flying squirrel from one tree to another, and found new and wilder territory where none but I ever went. I splashed along creeks where the waterway was the only means of advancing through an otherwise impenetrable forest of dense underbrush and stickers, and where the thick tangle of tree limbs and vines created a congested canopy which blocked out much of the light. Even darker were the crawlways through the thickets. These were animal trails which I found and widened, and which led to clearings at the bases of trees and which became my own private retreats where I was surrounded by the scents of sassafras and honeysuckle - and sometimes the odor of skunk cabbage. The outdoors became my refuge from the indifference I felt from the human world around me.

One winter in fifth grade I got into quite a bit of trouble when, instead of taking the bus directly home from school, Robert Cylinder and I slipped into the snow-covered forest behind the school yard and walked several miles to his house over Pennypack Creek, which was frozen solid due to extended sub-freezing temperatures. The trees were beautifully anointed with snow and ice, and the woods sparkled brightly despite the overcast: a prismatic crystalline fairy land. We were thrilled with the splendor and excited by the adventure. I called home from Cylinder's house in order to let my parents know that I would be a little late, as I had another couple of miles to walk. I never expected my father to be angry and to scold me for not taking the bus. Our spontaneous excursion did not meet with parental approval.

ON BECOMING A NONCONFORMIST

I discovered reading at an early age, and found that when I couldn't go outside to have my own adventures, I could experience them vicariously in the pages of a book. I was enthralled by the wonder and imaginative concepts of science fiction, and by the protagonists who explored alien planets and the far reaches of the Universe. I made up stories in my head, daydreamed extensively, and even turned a hand toward putting my stories on paper.

I was a nerd long before the word was invented. I dressed conservatively for school, stored my pens in a plastic liner in my shirt pocket, and carried my books on my back in a knapsack; the "in" thing was to secure school books with a rubber strap which had a metal hook at each end, and to tuck the books under one arm. The other kids made fun of me, but I didn't care. I didn't want ink stains on my shirt front, and I liked having both hands free and not having to worry about dropping my books all over the bus floor or crowded hallways. (In later years knapsacks became all the rage, so in retrospect it could be said that I was a boy ahead of my time.) I did what I thought made sense, and I felt no obligation to conform with manners or notions which seemed to me either wrong or irrational.

These traits in my character have caused constant frustration throughout my life. I was fanatically disturbed when my cousin Frankie Testa stole a nickel from me (this at a time and an age when a nickel meant something). We got into a loud argument over it, but no matter how much I cried and tried to take it back, he would not give it up. (He was older and bigger.) The screaming brought my mother into the fracas. When I stopped my tears long enough to explain what the fight was about, instead of confronting Frankie and seeking justice, she callously offered to give me a nickel in order to shut me up. I wouldn't take it. She was never able to understand that I didn't want *her* nickel, I wanted *my* nickel. And I did *not* want Frankie to have it.

I had quite an eye for spotting coins and tokens, a knack that others often found uncanny. My mother was always amazed when I bent down suddenly and retrieved a nickel or a dime or a quarter from the sidewalk. Why no one else saw them was a mystery to me: the coins appeared to lie in plain sight. I shrugged as I crammed my newfound gains into one of my overstuffed piggy banks.

Money Management

I always had money as a child. I sought out odd jobs, worked whenever I could, and was thrifty to the point of parsimony. I saved so much money that I opened my own savings account while still in grade school. I could account for the coming and going of every penny I ever had. One reason for my relative opulence was that I rarely spent my capital, choosing instead to let it grow with interest. And contrary to the philosophy of American society, I did not feel the need for material things.

In later years my father used to tease me that I still had my recess money. He was referring to the two cents he gave me to buy two pretzel sticks during morning recess. He claimed that I saved the pennies instead of buying the pretzels. I recall eating pretzels, and I have no conscious memory of not buying pretzels and pocketing the pennies instead. But it would not have been out of character for me to save the money and go hungry until lunch, as I did just that with my lunch money during college days when necessity demanded.

When I was seven or eight years old I was outgrowing my tricycle. My parents told me they couldn't afford a bicycle. So I earned and saved enough money to buy my own. I took care of it and rode it throughout my entire teenage years.

I mowed lawns, chopped weeds, pulled crabgrass, cut wood, shoveled snow, sold products door to door, collected soda bottles and scrap copper from construction sites, dug ditches, cleaned basements, ran errands, and so on.

Encounter with a Wild Dog

One errand found me biking to the grocery store to buy cigarettes for my parents. Along the way a mongrel dog raced out of an unfenced yard, barking angrily. Since it was common for untrained dogs to behave that way, I ignored it, but the dog surprised me by leaping and biting my leg. I pedaled hard to get away. At the grocery store I examined the damage: my pants leg was torn to pieces and my leg was scraped but not bleeding.

After making my purchase (minors were allowed to buy cigarettes in those days) I picked up a stout stick, laid it across the handle bar, and retraced my route, with the exception that instead of riding all the way back on the right side of the road as I normally would have done, I crossed over to the side the dog lived on as I approached the yard it was guarding. Just as I anticipated, the dog rushed out barking. When it leaped to bite my other leg, I raised the secreted stick and brought it down so hard on the dog's head that its jowl slammed hard against the unpaved shoulder. Ever since that day the dog barked furiously whenever I rode by, then, when it recognized me, it skidded to a halt and whimpered back to the house with its tail between its legs. We both learned a valuable lesson.

My Grandparents and a Second Home

I divided my summer breaks between Philadelphia and Salisbury. I went with my father to the construction sites and helped by doing electrical work. By age fourteen I could wire an entire house by myself with no supervision. In Salisbury I stayed with my grandparents, helped my grandfather paint, and did farm work, mostly tending the cornfield and picking watermelons for my great uncle Joe McAllister. This is not to imply that I didn't spend some of my spare time reading, or swimming and fishing in the local creeks.

Walter and Pauline Ruark were born prior to the turn of the century, before the invention of airplanes. My grandmother lived long enough to see man land on the Moon. They were wonderful people of the salt-of-the-earth variety. My grandfather was a self-employed house painter who also grew corn in the field next to their sprawling country homestead. They lived in a large, two-story house which my grandfather built. Also on the property were two garages (one for the pickup truck and paint cans and brushes, another, which had an attic, for the car), a chicken coop and hen house, a pen for a pair of hound dogs, several storage sheds, and a huge two-story barn which housed the farm implements, a buggy, a stall, and hay for the horse which spent the days in an adjacent corral. It was a great place to spend weekends and summer vacations.

They lived in true country fashion. My grandmother cooked on a six-foot-wide iron stove which was fueled by wood, and which made the kitchen the warmest room in the house and, consequently, in the winter, the gathering place. She pressed clothes with irons which were heated by resting them on the stove top. She gathered eggs in the morning for breakfast. She maintained a beautiful flower garden. And she prepared fried chicken from scratch.

I liked to collect eggs and delighted in chasing and catching the chickens for dinner, but I could never kill them. To my grandmother, a chicken was nothing more than food on legs. She held a conscripted chicken upside down by the legs, laid it across the chopping block despite its clucking and flapping, chopped off its head with one sure stroke of the ax, dunked the body into scalding water in order to soften the skin and loosen the feathers, plucked it bald, and cut it into pieces for the frying pan - all with the same nonchalance of ordering a meal at a restaurant. Hers was the best fried chicken I've ever tasted.

For you city folk, it's true: the beak does continue to open and close after the head is chopped off. And the headless chicken will sometimes stand up and run around in circles - but only for a few seconds. Bureaucrats can act like that for years.

My grandfather was my idol. He was a kind, good-hearted man who never hurt a soul, who never raised his voice in anger, and who never had an unkind word for anyone. I adored him for the values he stood for. When I was very young he took me for buggy rides on the winding dirt trails that cut through the vast pine forests and that forded the countryside's clear, pristine brooks. We picked huckleberries which my grandmother baked into pies.

He took me hunting: in the marsh for ducks and in the woods for rabbits. I was always quiet and kept my place behind him and his shotgun. One time I spotted a rabbit before he or the dogs took notice, and chased it away so it wouldn't get killed. My grandfather didn't get angry. Nor did he stop taking me on future hunts. Although I didn't care for the killing, whenever we came home with game I had no compunction against eating it.

My grandfather was a model of goodness. Before acting in response to a difficult ethical situation, I have always asked myself, "Would my grandfather approve of my action?"

The Painful Loss of Religion

I will always grieve over the anguish I caused my grandfather when I refused admission to the church by denying baptism. He was a Baptist deacon who attended church every Sunday, accompanied by my grandmother. My mother and I went to church whenever we were in Salisbury, but never in Philadelphia. My father was a Catholic who went to church only at Christmas and Easter; his confessions were very long.

My fall from grace and rise from innocence were directly attributed to Santa Claus. I was the last kid in my class to believe in jolly old Saint Nick. All the other kids taunted me with the truth I refused to accept. I clung tenaciously to my childlike convictions not because I was afraid of losing the illusion of the cherished gift giver and his flying reindeer, but because that truth undermined my ingenuous trust in the integrity of society. Admittedly, for years I had been uncomfortable with the authenticity of a fat man dressed in a red riding suit visiting millions of homes in a single night and carrying enough presents for all the world's children in a solitary sack, to say nothing of his miraculous aerial conveyance and remarkable chimney exploits. But I was willing to suspend my disbelief because of blind faith in the honesty of adulthood. My acknowledgment of the falsehood of the Santa Claus myth forced me to challenge the basis of *all* belief systems - not just children's fairy tales.

Once I realized that a belief was not founded on observable fact, but on an ideal that people wished fervently to be true, I became a realist. People have called me an atheist, but I am not. An atheist is one who does not believe in God. Since there is no God in which to disbelieve - except in people's dreams and imagination - I cannot disbelieve in God. At the tender age when I discarded belief in Santa Claus, I made a firm and conscious resolve to accept reality as it is, without creating fantasies for matters that I did not understand.

That was why I refused to accept baptism at the age of fourteen. Despite how much this hurt my grandfather, I could not go through the motions of baptism just to bring appeasement, for that would demonstrate contempt for the truth. My grandfather never let my loss of faith come between us. He loved me just the same.

Although I continued to attend church and Sunday school in Salisbury, I did so only in allegiance to family tradition. I regarded church-going simply as a conventional Sunday outing, as a chance to see and talk with long-time friends and relatives, as a social gathering followed by a big dinner prepared by my grandmother and by visits to or from other members of the family.

Not for a moment did I accept the fabulous story of creation as portrayed in the Holy Bible, or the revelations of miraculous events which were accepted without explanation, or omnipotent deities, or divine salvation. I accepted what could be proven or at least could be inferred. Yet I still have faith: in the innate goodness of the majority of people and in the eventual domination over man's baser motives.

Whole Hog pursuit

My best friend in Salisbury was (and is) my cousin Michael McAllister. My great uncle Joe was his grandfather. Michael lived next door to Uncle Joe, so he was actually raised on a farm and helped out with the chores. (Michael is now a veterinarian with a practice established on the old homestead). Uncle Joe was a full-time farmer with many acres of soy bean and watermelon as his primary crops, and a barn full of horses, cows, pigs, and chickens. Michael taught me much about farm life, such as how to drive a tractor and a stake-body truck long before we reached the legal driving age. We plowed fields together, planted seed, and picked watermelons in the dreadful heat

1950's: The Origin of Wreck-Diving - 81

of the August sun. It was good, hard work, and I loved it. At the end of a day of toil I felt as if I had accomplished something.

Michael and I had many adventures. Once we joined forces with Bruce and Rodney Adkins, brothers who lived in a shack about half a mile down the road from Michael, to explore the swamp and forest across a distant pond. We tramped through the woods to the point where the three of them had built a raft out of a couple of fifty-five gallon drums which were overlaid with planks. We poled across the pond into what was to us unknown territory.

After landfall, we prowled along the water's edge till we came to a wooden fence that went all the way into the pond. That seemed odd. We didn't know what to make of it, but that did not stop us from climbing over the top to see what was on the other side. Some minutes later we heard a deep-throated grunt behind us. At first we couldn't see anything because the underbrush was so thick. The grunting continued, louder, and what finally strutted into view was about a six hundred pound and very ornery hog whose territory we had just invaded.

This hog wasn't just angry, it was mad. It charged, and we ran like the dickens. To say that I was terrified would be a gross understatement. I fled as if the devil itself were nipping at my heels, as it almost certainly was, judging by the wild stentorian snorts that were continuously erupting from all-too-close behind. The hog was nearly upon us when we encountered another fence. We were doomed.

Older brother Bruce vaulted over the fence in a single bound; only his hand momentarily touched the top rail. I couldn't jump that high, so I got one foot on a lower rail, pivoted on my belly on the top rail, and swung over gracelessly to land on my back in the weeds. I don't know how Michael or Rodney got over, or whether they preceded me or not, because we each hit the fence at a different location and I was too absorbed with saving my own life to take careful observation. The hog ran back and forth along the inside of the fence, snorting fitfully only inches from my face. I crawled away to where the others had gathered. We were all shaken up by the episode. But another predicament soon presented itself: how were we going to get back across the pond with an angry hog guarding our return route to the raft?

This fence also stretched all the way into the water, apparently so the hog could wallow in the shallows as hogs are wont to do. We followed the fence up a gentle slope and through dense brush and thickets. The hog kept apace, now snorting thunderously in impotent rage. Eventually, the primeval forest terminated at the crest of a rolling hill; beyond lay a cleared field which seemed to go on forever, as did the fence that marked the confines of the hog's gigantic pen. We agreed on a plan. We walked along the tree line in plain sight of the angry beast, ducked into the forest when we hoped we were far enough away, trampled through the underbrush till we regained the water's edge, then sneaked clandestinely to the fence. We listened carefully to the birds chirping in the trees, the frogs croaking in the marsh, and the insects buzzing through the air. We exchanged silent nods. I took a deep breath, then scrambled over the fence alongside the others and ran like mad before the hog caught on to our strategy. We clambered over the other fence long before the hog knew we had escaped.

PLAYING WITH TOY GUNS

Michael taught me how to shoot an air rifle (or pellet gun) and a BB gun. He protected the fields by firing pellets or BB's at birds that were gobbling up the newly sown seed. The distance was so great that he seldom if ever hit the feathered filchers, but the puffs of dust as the shot struck the sandy soil were enough to frighten off the flocks. We practiced on more mundane targets such as cans and bottles and cardboard cutouts.

One time he placed an old shaving cream can on a picket fence, and to make the target more difficult to strike he laid it on its side so only the bottom pointed our way. Michael's marksmanship produced an unforeseen consequence. Instead of glancing off the metal, the BB punctured the concave base and explosively released the remaining propellant and cream - all over one of Aunt May's prize bushes. We spent the next half hour hosing down the bush and wiping off each individual leaf before she discovered the desecration.

Stalking turtles was one of our favorite summer pastimes. The pond was filled with partially sunken logs on which the cold-blooded reptiles sunned. Because every turtle slunk into the water at the barest hint of movement, we crawled on our bellies to a hidden position from which we could aim our sights at the quarry. The BB's plinked harmlessly off the hardened shells. Once hit, the target turtle took refuge at the bottom of the pond but the nearby turtles remained. The trick was to shoot fast enough and accurate enough to hit the same turtle twice before it dropped off the log and submerged. One slow or dimwitted turtle took its time getting into the water, and I hit it three times consecutively. I was so exuberant at my performance that a moment later, when a bird chirped on a high tree limb behind me, I swung the gun and fired without thinking. The BB went right through its narrow neck, and it fell to the ground, dead.

The horror of my thoughtless action gripped me instantly. I burst into tears and sobbed uncontrollably, all out of proportion to the event. Michael could not understand my reaction or perceive my inner turmoil. He tried his best to console me: after all, he said, it was only a bird. But whereas birds are often considered nuisances, or at best a lower life-form which is tolerated by man, the picture that came to my mind was that of a nest that would be without a protector, of the chicks that had lost a mother or father, or the parent whose mate would never return. Perhaps I was an overly sensitive child. (I was careful not to step on anthills so as not to squash any ants.) Perhaps I ascribed human feelings to nonsentient creatures in which emotions do not exist. Or perhaps the incident forced me to confront some deep subconscious insecurity. I threw down the BB gun and never fired it again.

THE BOY SCOUTS

At home I joined the Boy Scouts because of the opportunities that were offered to gain woodcraft knowledge and experience. I met other boys with similar interests, none of whom went to my school or lived in my neighborhood. The simplistic training program was a worthwhile endeavor, although it could have been more influential had our troop had better leadership.

With membership came hikes, jamborees, and overnight camping trips in upstate backwoods. Since my parents were interested only in social activities and had no desire to tramp through the timberland or to spend a night in the cold, the Boy Scouts was my only structured outdoor outlet. It was a learning activity that complemented my personality. I loved the world of nature: not as much from a scientific standpoint or because of esthetic appeal as from the thrill of discovery and, later, the physical challenge.

As patrol leader I organized hikes into the nearby woods. I soon found that we could not cover much territory in a single day, and we were not equipped for overnight backpacking. None of us owned sleeping bags, backpacks, or cook stoves, nor had any prospect of obtaining them. But since we all owned bicycles, I came up with the idea of going on bike hikes. With a packed lunch and a full canteen we roamed many miles through the woods, traveling farther than we ever could before and reaching areas that were new to us.

We bushwhacked across untrammeled ground because there were no trails. When the terrain became too rough for riding, we dismounted and pushed the bikes till we reached more tractable ground. We rode fast across shallow creeks without getting our shoes wet. We coasted down steep hills, controlling our speed not only with the brakes but by picking the appropriate angle of descent. Today much of those woods have become a park, and I jog in an hour what was then an all-day excursion.

ADVENTURES UNDERGROUND, OR EXPLORING THE SEWERS

My greatest hometown adventures in the late 1950's took place in the city's sewers; or, to be more precise, the storm drains. Sewers are so called because the pipes drain sewage from toilets and sinks in the house, whereas storm drains empty the streets of rainfall and melting snow. They are two separate systems with no interconnections. There is no sewage or human waste in the storm drains. As kids we referred to the storm drains as sewers, so I will maintain the convention here.

Concrete sewer pipes are often placed in the beds of naturally occurring streams: channels eroded in the earth over millennia and which draw water off the land and convey it to rivers which carry the water out to sea. When roads are paved or buildings are constructed over the courses of regular flow, provision must be made for the drainage of water from higher ground to lower. Thus a sewer (or storm drain) can be visualized simply as an artificially covered creek. This intricate network of underground passages is equivalent to a limestone cavern system, but with much more order in its arrangement.

The first "sewer" I entered as a child was an arched brick tunnel that permitted the creek at the end of my street to flow unobstructed under the raised railroad bed. It was no longer than twenty-five feet, and high enough so I could stand with room to spare. Both ends were veiled behind wild tangles of brush and vine. The rocky bottom was about ankle deep. The interior walls were overlain with webs whose spinners survived by devouring the bugs that flew or crawled inside. Calcium stalactites dripped from the ceiling the same as they do in caves.

Once I got over my fear of spiders and darkness, the tunnel became an attractive retreat. The temperature inside was cool in summer, warm in winter. There wasn't a whole lot to explore, but the tunnel gave me notions of similar conduits that I found under streets in the neighborhood. I sought sewer outlets wherever I could find them, and crawled far inside each one until the entrance behind me appeared only as a tiny circle of light. Stygian blackness lay ahead. To go any farther required some form of illumination.

The household flashlight allowed me to extend my explorations, but the beam was as dull as it was uncertain. I found an abandoned signal lantern by the railroad tracks, repaired it and bought wicks and a gallon of kerosene, and replaced the red lens with one that was clear. This gave me the prolonged light I needed to probe the sewers more fully. The flashlight became my back-up. The flashlight was handier and more effective in the small pipes in which I had to crawl, but these side pipes eventually led to the mains in which I could stand up straight and walk. The mains were rectangular and were made of reinforced concrete that was poured over forms which were built on location, and which were removed after the concrete set. The side pipes were tubular prefabricated concrete sections which were buried in the ground end to end. Picture the entire network as a large rectangular tunnel from which smaller tubular pipes branched, or tubular tributary pipes feeding into a larger rectangular main.

Most of my underground excursions I carried out alone. Occasionally I talked some of the neighborhood kids into going with me for short outings, but my primary companion on many of the longer jaunts was my younger next door neighbor, Tom Gmitter, now an optometrist and still my good friend although geography has decreased our contact.

Over the years I explored many offshoots of northeast Philadelphia's sewers. I developed routes by which I could travel for miles without surfacing, or, if I did break out into sunlight, I did so at the edge of undeveloped land at which point the outlet poured into a creek which then meandered through a deserted wooded area and eventually led me back underground. I found several ways of passing under the six-lane Roosevelt Boulevard. One sewer traversed beneath a large shopping mall and its parking lot. Another enabled me to sneak under the perimeter fences of the airport and emerge in the scrub, from which I had access to the grounds and runways. I watched airplanes take off and land, sometimes at night. I often spent hours crawling up shrinking side pipes until they became so small in diameter that further progress was prevented. If I could not reach an overhead shaft in which to climb and turn around, I had to back out. At the farthest point of these journeys I tried to lift the manhole covers in order to learn where the pipe had taken me. More often than not, the covers had not been removed for so long that they were wedged in place with rust or dirt and could not be moved with the leverage available from standing on the ladder rungs and pushing up with my back.

The method of walking through the tubular sewers was a variation of the waddle. In order to avoid the trickle of water that ran down the middle of the pipe, we stepped high on the inward curving sides. This kept our shoes dry. Running in this fashion produced an awkward, syncopated gait. The higher the water, the more pronounced the waddle, until it reached the height of inanity, especially if the pipe was less than head high and we had to duck (pardon the pun) as we walked. Flood waters produced by tremendous rain storms kept the pipes swept clear of debris.

The broad rectangular mains developed bottoms very much like shallow creeks. Sand and dirt washed in from the streets formed dunes, sometimes packed with beachlike consistency, sometimes coagulated like caked mud or clay. The water course meandered from side to side the same as a stream in the woods. We were usually able to walk on the hardpack instead of wading through water.

The mains ranged in size from eight to ten feet in height and width, to as much as twenty feet in both dimensions. When drastic changes in elevation were demanded by the terrain, long steps like salmon ladders were constructed, creating waterfalls. There was an entire ecosystem living underground, largely on the insect level. In all my years exploring the sewers I saw but a single rat, and it scurried away as I approached.

As in any form of exploration, not every trek went smoothly. The worst scare I had in the sewers occurred during a solo excursion. I found a tubular outlet which angled down into the ground at about thirty degrees. The pipe was less than two feet in diameter. I crept in head first by sliding and pulling myself forward on my elbows. After proceeding a couple of body lengths in the head-down position, I wondered how difficult it would be to back out (and up) if the pipe were obstructed farther down. In the beam of my small flashlight I couldn't see any turn to the horizontal, which meant that either I still had a long way to go to reach a bend, or that the pipe came out near the top of the connecting conduit. Landing on my head on a concrete slab from a height could be painful. I stopped and backed up with difficulty, and the going was slow and arduous. I had to roll my hips from side to side and scoot back with my elbows because there was not enough height to crawl. I was frightened by my slow progress and by how much energy I was expending. When I got out I was breathing hard.

After I caught my breath, I turned around and crept down feet first. The descent was deceptively easy. After about thirty feet I came to a crook where the pipe leveled out. I rolled onto my back and wormed slowly past the bend. Again my beam of light failed to penetrate far enough to enable me to determine what lay ahead. I didn't like the uncertainty of my position, so I proceeded to slither back up. The going was easier but exhausting: I moved forward by inches, and had to wedge myself against the sides between progressions so I did not slip back down. I thought that by forcing my legs under me I could crawl and make better time while conserving my quickly depleting strength. I lay on my side and tugged on my leg with my hands. My knee stuck at first, but by forcing it into the exact center of the pipe where the diameter was the greatest, my leg suddenly popped forward -- and lodged me firmly in the pipe with my left leg jammed against my chest and my right leg stretched out behind me. The effect was like that of forcing a knuckle into the narrow neck of a bottle. I couldn't move.

I was stricken with panic. I struggled like a madman in a straitjacket. No one knew where I was, and I was not likely to be found in so remote a place. My parents paid no attention to my activities and would not come looking for me. The rush of adrenaline gave me uncommon strength. I strained until I tore my pants and several layers of skin off my knee, and finally succeeded in unbending my leg. I crept up the pipe and into the glorious sunlight.

After I calmed down, I paced out the distance to another culvert a block or so away. (To a Philadelphian, a city block is one tenth of a mile, or a little over five hundred feet.) I re-entered the sewer main, then counted my steps past side pipes till I found one of the same diameter at the approximate distance from my starting point as the one in which I had been trapped. I crept into it, came to the familiar bend, spotted the circular glow of light above, and crawled out. That tube then became one of my secret entrances.

On one lengthy excursion, Tom and I were a long way from the entrance or any intermediate exits when the lantern ran out of wick about the same time that his flashlight batteries died. We had to feel our way out in the dark. This is not as difficult or as scary as it may sound. Sewers are not like labyrinths or convoluted mazes. Downstream always leads out. We simply waddled through the pipe till we reached the junction of the rectangular main, then turned right and groped along the concrete wall. We didn't have to go more than a quarter of a mile. Another time I ran out of kerosene. After that I carried spare wicks and extra kerosene, and, by practicing in the basement with the lights out, learned to change the wick and refill the reservoir in the dark.

The day that a terrible downpour inundated the city, Tom's father, remembering that Tom had taken the flashlight and gone out with me for an afternoon's diversion, knew instinctively what our intentions were. He proceeded immediately to the sewer main entrance we used most often, behind the shopping mall. He stood there in galoshes and raincoat with water cascading off his southwester, and beheld in horror the raging flood that filled the seven-foot-high sewer to the very top. He feared the worst. He prayed that instead of being drowned we had retreated up a side pipe or climbed up into a shaft and either escaped through the manhole cover or, if the cover was stuck, could cling to the rungs until the flood subsided. He returned home to wait out the suspense.

Earlier, when Tom and I arrived at the entrance and saw the muddy water that was flowing already ankle deep, we surmised that the darkened sky portended that the light drizzle might turn into something more serious. We decided to go climb trees instead. When the mighty rains arrived, we stashed our stuff in the woods and spent the rest of the day in the mall. His father was relieved when we arrived home safely.

No Foreshadowing

The thought of scuba never entered my mind. Like most kids, I played in swimming pools, swam in creeks, and surfed in the waves on an air mattress. I even bought a face mask which I wore several times in ponds but only once in the ocean - a comber ripped it off my head and I never saw it again. Nor did I bother to replace it.

I once owned a snorkel. It was the kind designed for children. The top had a gooseneck which curved down and terminated in a cage which held a plastic pingpong ball. Upon submergence, the ball floated up and sealed the opening so that water could not fill the tube. I never learned how to breathe through it.

I detested *Sea Hunt*, the television show that starred Lloyd Bridges. The story lines were stupid, the fighting scenes were implausible, and the continuity was badly flawed. I didn't care for boats or trips across the ocean. The only ships I dreamt about were spaceships.

At a time when most boys wanted to be firemen or policemen when they grew up, I wanted to be a trash collector. I wasn't thinking about the security of a job in which there would always be work. I simply liked reclaiming articles that people discarded or no longer wanted, and finding further use for them. I was into recycling long before the concept became a national initiative.

To me it seemed socially derelict for people to disown perfectly serviceable items because they got tired of having them around, or because they wanted newer products in order to advertise their wealth. I was too young to read about conspicuous consumption, but I knew it when I saw it. Waste was an affectation of affluence - a concept which I staunchly repudiated.

There were other occupations that interested me as I grew older, but the opportunity to pursue them was not afforded to me. Let the decade of innocence end, and a new decade of truth and tragedy begin.

Part 4
1960's: Deep-Water Triumph and Turmoil

Wreck-divers Go to Sea

The big break for wreck-diving came when Mike de Camp found Joe Galluccio, who ran the charter boat *Sea Ranger* out of Brielle, New Jersey. At that time the *Sea Ranger* operated as a fishing boat, transporting people to offshore spots where fish were abundant enough that a person could go home with a satisfying catch and food for a few meals. The two types of marine sport angling are trolling and bottom fishing. In bottom fishing, boat captains seek out "holes" and ridge lines where fish tend to congregate. The captain who has the greatest knowledge of spots that consistently produce fish can count on more business than less knowledgeable competitors.

The ocean bottom off the New Jersey and adjacent coasts is a flat, featureless plain: a giant sandbox very much like a desert and unlike the tropical coral reefs found farther south, or the kelp-covered seabed indigenous to the north and along much of the west coast. Plenty of fish inhabit these underwater grounds, but they are spread out fairly evenly over a large geographic area. Besides the concentrations noted above, another place where fish gather in abundance is on shipwrecks, whose large mounds of wood or metal provide a substrate for the smaller organisms on which fish feed, and whose collapsing hulls offer bottom dwellers places to hide from predators.

A shipwreck is an artificial reef. Fisherfolk have known this longer than anyone.

Finding Shipwrecks

At first, the only way one could return to a known site was by taking note of land ranges. In this method of relocation, the position of a hole or wreck is pinpointed by triangulating at least two features on shore, such as a telephone pole behind the chimney of a house with blue shutters, and a water tower standing at 265 degrees. It takes a fair amount of experience to develop this navigational skill and, as in every other endeavor which requires expertise, some people are better at it than others. Rain or fog may obscure the landmarks, and the horizon presents an impenetrable barrier which limits the triangulation method to wrecks which lie near shore.

Galluccio had an advantage over most other charter boats in that the *Sea Ranger* was equipped with loran (acronym for "long range navigation"). This is an electronic system of triangulation which the Allies developed during World War Two. Transmitting stations on shore send signals to a receiving unit on a ship or boat. By aligning at least two signals, the operator can determine his true position. Not only is the accuracy greater than landmark triangulation, but the system works at night, under most weather conditions, and reaches several hundred miles at sea. Once a wreck was located and its position recorded, one could theoretically return to the same spot at any time.

Many problems were encountered with loran A, as the early American version was called. Sometimes the transmitters were down or the signals were out of phase. In the latter case one could search for hours without locating the wreck because the transmitted reading was false. I once spent an entire day in a wetsuit thinking that the boat captain would locate the wreck at any moment. We never got to dive, and I was seriously overheated and dehydrated by the time we called it quits and returned to the dock more than frustrated.

Primitive loran receivers were not as accurate or as easy to operate as modern units. Instead of displaying a six digit code with decimal precision or the latitude and longitude determined to the second, an oscilloscope flashed two rows of vertically spiked lines which were constantly in motion: like fluctuating EKG's. The operator dialed in numbers which represented the time delay (TD) of each transmitter's signal to reach the place where he wanted to go. The trick was to align the two series of spikes so they matched perfectly, this being the point of convergence of the two incoming signals and, theoretically, the sought-for location. It was easy to misinterpret the alignment which occurred at the point of intersection.

Galluccio overcame the deficiencies of loran by installing side-scan sonar (acronym for "sound navigation and ranging"). This is equivalent to an underwater radar unit except that instead of bouncing high-frequency radio pulses off distant objects in the air, sonar bounces sound waves off distant objects in the water. The resultant detection of a shipwreck on the bottom returns a ping the same as that heard in movies in which a destroyer tries to find a submerged enemy submarine. This is also called echo location. Several times I've been in the water when the boat captain had forgotten to switch off the sonar unit after the commencement of diving. I felt my lips tingling or crawling with weird pulsations whenever I faced the boat. The rest of my body was insulated from the pinging by my neoprene rubber suit. Sonar also has its shortcomings: the signal is sometimes reflected off the thermocline (the division between layers of warm and cold water) and returns no information, the same as it would if there were no wreck in the vicinity.

Fishing Charter Tag-a-longs

Another boat that played a part in the early days of wreck-diving was the *Thumper*, Captain Teddy Weeks, which operated out of Barnegat. Weeks owned a British loran unit called Decca, which was more reliable than American loran, and he was proficient in using it. Later came the *Big Jim*, Captain Jim Dulinski, out of Cape May; *Albatross III*, Captain John Shukus, out of Barnegat; *Captain Chum*, Captain Chum Robbins, out of Stone Harbor. Operating from Long Island ports were the *Jess-Lu*, Captain Jay Porter, out of Freeport; and the *Viking Starlite*, Captain Paul Forsburg, out of Montauk. De Camp remembers that Porter could find wrecks better than anyone.

Since these boats maintained a full fishing schedule for the summer season, they were generally available to divers only in the winter months, when the fishing business was off and the temperature of the water hovered barely above freezing. Divers had to be tough to stick with the activity in those formative years. They went out all day and shivered in a cold cabin till the anglers brought up their catch, then made a single dive that was ten or fifteen minutes long. Decompression times usually did not exceed ten minutes.

When dive boats came into their own, one could add perhaps five minutes to the dive, then, after waiting for two or three hours for residual nitrogen to escape, one might make a repetitive dive for as long as ten minutes. Bottom times of such duration seem short by today's standards, but remember that present day divers have the benefit of better equipment and greater knowledge not available to wreck-diving pioneers in the 1960's.

It should be understood that in the beginning these party boats were not equipped with ladders. This made it difficult for divers to climb back aboard after a dive. The usual tactic was to hold onto a rope tied off to the rail, wait for someone (or two) to grab the tanks, release the straps and shrug out of the iron encumbrances, have the tanks hauled on board, then climb over the tall gunwale with assistance from above while kicking like mad. Good physical conditioning helped.

Clandestine Wreck-diving

Then, as today, nearly all shipwrecks were initially found by draggers or trawlers, whose fishing nets hung up on unknown obstructions and often could not be freed, resulting in the loss of some very expensive gear. In order to avoid future losses, these commercial fisherfolk recorded the position of the site so they could stay away from it. Practically every wreck I have ever dived had at least one net snagged in its broken-down hull or debris field; some wrecks have several nets in various stages of decay, representing years of accidental relocation.

Nets and fishing line are the primary reasons that wreck-divers began carrying knives.

Every dragger and trawler has a book of numbers in the wheel house: a list of obstructions found by them or their associates. Head boat captains (and later, dive boat captains) curried the favor of these commercial fisherfolk in order to have a look at their black book and come away with some very hot numbers. Once obtained, these numbers became a closely guarded secret which were not shared with anyone so that a charter boat captain could take his customers to places where no one else could: secret spots that his competitors did not know about. This state of affairs has persisted to the present day. Indeed, violent arguments have erupted between boat captains on the suspicion that someone has peeked into his little black book and stolen his private numbers, or when a boat was caught grappled in a little-known wreck with divers in the water while a competitor who previously did not have the wreck's numbers passed by close enough to read them off his own loran. Sometimes the subsequent arguments became physical, but more often they developed into abusive confrontations and lifelong animosity.

Although it was not his regular business (or anyone's regular business, for that matter), Galluccio was willing to take de Camp and his fellow divers to offshore wreck sites in order to dive instead of to fish, but only in the off-season. For that reason, the group sometimes referred to themselves jokingly as the Winter Time Divers. Galluccio had compiled a list of loran numbers for wrecks which had never been dived. Most were unidentified, but some were known by name because their positions had been marked on the charts as one-time hazards to navigation or because they were torpedoed by German U-boats during the wars and had been buoyed for the purpose of potential salvage.

De Camp and about fifteen loosely knit divers began exploring the larger wrecks that lay beyond the sight of land. De Camp did the organizing. He obtained the various boats to go to specific locations, then telephoned the divers and got them to commit to available dates. To take up some of the slack during the summer, Jack Brown, a trucking company owner and fellow diver, took a few people out on his 30-foot sportfisherman *Daisy B*.

Primitive Equipment

Because these offshore wrecks were sunk in deeper water, dive gear was modified to meet the conditions. Two tanks were banded together to form twin tanks or "doubles." A "cheater bar" consisting of a valve stem in the middle of a length of tubing which to the valves of both tanks completed the ensemble. The double-hose regulator was connected to the central stem, then each tank valve was opened so that all the air was directed to the rubber mouthpiece.

Submersible pressure gauges were unknown at the time, so a diver never knew how much air he had remaining. Eventually came the J-valve, a device which held back some of the air (300 to 500 psi) until a rod was pulled, whereupon the rest of the air was made accessible while the diver ascended to the surface. There were serious uncertainties with the J-valve. The valve too often got knocked down accidentally, so when breathing became difficult and the diver went to pull down the rod, he found no reserve air supply waiting for him. Now his last breath was only seconds away, so he had to spurt for the surface like a nuclear missile launched from an atomic submarine. If he did not have enough air to make it, he drowned or suffocated. If he panicked and held his breath, he could die a horrible death of embolism as the expanding air burst his lungs like an overblown balloon. If he stayed calm and reached the surface alive, he might incur the paralyzing effects of the bends from too rapid an ascent. For one planning to stay down long enough to incur the *need* for decompression, a pulled-down rod meant that even if he made a proper ascent, there was not enough air to conduct needed decompression.

The Hazards of Decompression

Nearly as bad was the customary procedure used for staged decompression when a diver could not relocate the anchor line due to disorientation or to dark or turbid bottom conditions: he surfaced from wherever he happened to be when he got low on air, swam to the boat, and went back down the anchor line to the appropriate depth. Add to this the general lack of understanding among recreational divers of the absolute necessity for adequate decompression, poor appreciation for the harmful consequences of the bends, and little or no knowledge of its symptoms or how to recognize them, and deep, decompression diving was a dangerous activity indeed.

The wonder of those primeval times was that there were not more accidents than there were - likely due to the small number of people who were involved in the overall sport, to the quality of those who sought deeper depths, and to the brevity of the deep dives conducted and the short decompressions that were required of them, generally in the ten to fifteen minute range.

The problems of decompression diving were further exacerbated by the nonavailability of depth gauges and waterproof timepieces. The bottom depth was taken from the charter boat's depth recorder. Decompression stops were "eyeballed." De Camp's first underwater timepiece was a Timex watch housed in an aluminum cup with a clear plastic top, which he admits he was sometimes too "narked" to read in very deep water.

Before BC

The best thermal protection available was a 1/4-inch neoprene wetsuit consisting of close-fitting pants, jacket, hood, gloves, and bootees. Buoyancy compensators were non-existent: there were no lead weightbelts or inflatable life vests.

A diver floated on the surface like a bloated seal, pulled himself vigorously down the anchor line until the pressure of the water compressed the neoprene to the point where his buoyancy became negative, then he fell to the bottom and crawled around the wreck like a crab or swam in an upward angled posture while kicking hard with his fins. To climb up the anchor line meant hauling hand-over-hand until positive buoyancy was attained, at which point ascending became easier.

To ascend without the anchor line required kicking off the bottom with powerful strokes that demanded strength and stamina and that consumed a considerable amount of air. When the buoyancy turn-around point was reached, the diver threw up his hands in order to decrease speed during the remainder of the ascent. Thus the reason for swimming to the anchor line on the surface and going back down in order to satisfy decompression obligations. If all this sounds a lot like Russian roulette, that is an opinion which you should hold in check until you learn what future generations say about current methods. You go with what you've got.

The Founding of Shipwreck Research

New York harbor is the most active port on the North American continent, and has been for centuries. The coasts of New Jersey and Long Island create a funnel which concentrates shipping and which multiplies the potential for shipwrecking. With this rich area as their hunting ground, early wreck divers descended upon numerous wrecks that had been fished for years but never before seen by the eye of man.

De Camp and his fellow divers began the process of identifying shipwrecks by correlating known sites with historical data (a method which requires extensive archival research) and by recovering parts of wrecks on which lettering either spelled out the ship's name or gave enough information to lead the persistent researcher to its identity. Some wrecks took years to identify, some are still unidentified, some have yet to be found. The process is ongoing.

Oceanographic Historical Research Society

Another group, consisting principally of New York residents, entered the domain of wreck-diving and research in the early 1960's. They called themselves the Oceanographic Historical Research

1960's: Deep-Water Triumph and Turmoil - 95

Society, an appellation which was pretentious rather than prestigious and authoritative, and which might mislead one to believe that it held approved scientific standing. The OHRS had no official endorsement or institutional affiliation. It was a private club comprising a dozen or so members.

The nominal leader of OHRS was Graham Snediker, an air-conditioning engineer. The OHRS was considered a rival group by those who were residents of New Jersey. De Camp was voted into the club but, despite the opportunities that membership offered, he found that diving with the group was restrictive and unsatisfying.

De Camp felt that the OHRS was overly regimented, diving "by the numbers" one could say, as each diver was assigned a partner, a place on the boat for gear stowage, and an order for entry. Only one pair of divers was allowed in the water at any one time. The pair designated for the day as team number two couldn't dive until team number one returned to the boat. Team number three couldn't go in until team number two was out of the water. No one was permitted to dive alone.

Using two six-pack boats simultaneously - the *Three Daughters*, Captain Charlie Fischette, and the *Beaujean* - helped alleviate the waiting time, but these boats were available predominately in winter when sport fishing charters were scarce. De Camp was used to a more relaxed and freeform spirit.

Despite these drawbacks perceived by Mike de Camp, the OHRS conducted valuable research on wrecks off the Long Island coast, in particular the *Oregon*, a British passenger liner which sank in 1886 after collision with a schooner reputed to be the *Charles H. Morse*. Snediker and Charles Dunn recovered the first of many china dinner plates which provided the key to the wreck's identity - they were stamped "The Cunard Steamship Company" and carried the shipping line logo. The OHRS did not survive the decade.

SHIPWRECKS IN THE MAKING

Some noteworthy events which occurred in the early 1960's had a direct and profound impact on later wreck diving. One was the sinking of the Dutch freighter *Pinta* in 1963, after collision with the *City of Perth*. The publication of de Camp's startlingly clear photographs of this pristine shipwreck lying on its side in 90 feet of water did much to capture the imagery of what wreck-diving was all about.

The *Pinta* casualty was succeeded the following year when the Norwegian tanker *Stolt Dagali* was cut in two by the Israeli liner *Shalom*. The bow section of the *Stolt Dagali*, consisting of the wheel house and tank compartments, remained afloat and was towed to a New York dry dock. The stern section sank, leaving the diesel engine room and crew's quarters on the bottom at a depth of 130 feet: a prime target for wreck explorers who dared the unknown at a time when 130 feet was considered deep.

Both these wrecks lay within easy reach of boats out of New Jersey, and both were intact. They became lures for divers who were drawn to the darkened interiors just as moths are drawn to light. Later, when Ed Waverly lost his way inside the *Stolt Dagali* and died, the diving community earned a greater respect for the hazards of penetrating these black, murky corridors.

Public attention was focused on wreck-divers as valuable adjuncts of society whose expertise could prove useful. De Camp and friends descended to the *Stolt Dagali* just three days after it sank. They found the wreck by following the oil slick to a dark pool of floating petroleum. During his exploration of the mammoth hull, de Camp peered into an open porthole and spotted the body of a stewardess who was pinned to the overhead by the buoyancy of her life vest. He reported the gruesome discovery to the Norwegian embassy. Not only were the Norwegians overwhelmingly thankful for the information, but they asked if it was possible to recover the body and to search for others that might still be trapped inside the hull. De Camp said yes.

About three weeks later, de Camp led a team of divers to the *Stolt Dagali* with just those purposes in mind. Working together, the group first pried open the jammed door to the laundry room with a crowbar, and after great exertion removed the body of the stewardess from inside the wreck. Her feet and face were gone, and loose flesh sloughed off against the divers' wetsuits as they squeezed the body through the doorway, brought it to the surface, and stuffed it into a body bag provided by the Norwegian embassy. An undertaker met the boat at the dock and took charge of further transportation. No other bodies were found.

First Descents

De Camp did not publicize these grisly events, but he wrote articles about other shipwrecks he explored. These articles dramatized the lore and lure of wrecks and wreck-diving, and injected enthusiasm into others who had a similar bent for adventure into the unknown. In print he cautioned those who might enter the activity unprepared, "When shedding warm clothes to get into an icy suit it is sometimes felt that although we don't have to be crazy, sometimes it would help."

De Camp recorded a remarkable number of first descents - not that he organized every trip on which divers descended to wrecks that had never been seen before, but he was at least one of the handful who can claim to have dived the wreck when it was still a virgin. His inventory of first descents reads like a litany of today's wreck-diver's list of sites to see: *Cherokee, Coimbra, Durley Chine* (known at the time as the Bacardi and not identified until a quarter of a century later), *Hvoslef, Ioannis P. Goulandris, Lillian, Moonstone, Northern Pacific, R.P. Resor, Texas Tower, Tolten, Varanger,* and others.

His influence on the next generation of wreck-divers was pronounced.

The First Book on Local Wrecks

Also notable in the annals of wreck-diving was the publication in 1965 of a slender volume titled *Shipwrecks off the New Jersey Coast*, by Walter and Richard Krotee. It was the first book of a kind that catalogued the most well-known wrecks sunk in the area. Each entry was annotated with a brief historical synopsis - often only a sentence or two - and in a few cases included a description of the wreck as it lay on the bottom. The book proved an instant success and ran to two more editions, one in 1966 and the other in 1968.

The gross inaccuracy of the information contained in the book makes it appear to have been written without reference to published documents, despite acknowledgment of official sources the authors knew to exist - as if the Krotees relied primarily on local legend instead of ascertainable facts. Yet the thin misinformative accounts stirred a fervor among wreck-divers the same as the previous generation of underwater treasure seekers was stirred by Harry Rieseberg's fabulous hard-hat exploits that he promoted as truth instead of grossly elaborate hoax. If the Krotees failed to compile a work of indisputable fact, they succeeded admirably in heightening awareness among potential wreck-diving enthusiasts.

A Lasting Partnership

It was over the *Stolt Dagali* that Mike de Camp met the recently certified George Hoffman, an elevator constructor who was diving off another boat. During surface interval, the divers from the two boats gossiped over the gunwales. That night de Camp and Hoffman met at the dock and discussed their mutual fascination for shipwrecks, and a rapport was made.

Hoffman was so entranced by wreck-diving that he bought a 16-foot Boston whaler with an outboard motor. He and de Camp used the boat to dive near-shore wrecks off Brielle when no commercial charters were available. But the deep-water wrecks continued to provide the main attraction for them and for the coterie of divers with whom they associated.

The Mud Hole Wrecks

The main obstacle presented to deep wreck-diving was the geography of the sea bed. The bottom off most of the east coast of the United States deepens gradually to the east. Within swimming distance from the New Jersey shore the water attains a depth of about 30 feet. At a distance of thirty miles the bottom is found at about 130 feet, with some slightly deeper troughs occasionally intruding. An incline of 100 feet in thirty miles is not perceptible to the human eye - the sea bed appears to be flat. At sixty miles from shore the bottom is generally no deeper than 190 feet. This translates to long boat rides to reach the 130-foot wrecks (except for the *Stolt Dagali*, which lies in a trough only twenty miles from shore) and little opportunity of going farther and deeper due to daily time constraints and the slow speed of the average party boat.

Fortunately for the evolution of deep wreck-diving, a training ground was available that lay not only close to shore, but close to the population centers that supplied the growing number of divers who were engaging the activity. This training ground is a deep channel which has been gouged out of the bottom by the outflow of the Hudson River. The channel issues from New York harbor and trends approximately southeast for more than a hundred miles, at which point it deepens and is known as the Hudson Canyon, which lies near the edge of the continental shelf.

Due to the discharge of industrial waste and the dumping of sewage and drainage from congested metropolitan areas along the river, the water that runs through this channel is unusually dark and turbid, generally restricting visibility to only a few feet. Particulate matter predominates, hanging like a black pall over the wrecks which came to rest here. Thick silt has accumulated on rusted decks and inside inky rooms. This silt is stirred by the slightest touch or fin kick into swirling ebony clouds.

This murky channel of everlasting eclipse is called the Mud Hole.

To state that diving in the Mud Hole is a challenging endeavor is a foolish understatement indeed. Just as vegetation grows thick along the banks of a desert stream, so the constant supply of nutrients from the Hudson River feeds a thriving marine environment. More than in the nearby underwater plains, a ship sunk in the Mud Hole quickly becomes a lush, life-encrusted wreck, covered with filter feeding organisms on which small fish feast. Larger fish devour the smaller and so on up the food chain until man appears at the top, dropping baited hooks to catch the game he likes to eat.

Some anglers snag the wreck with their rigs instead of catching their intended prey. The lines are cut at the surface and the heavy-duty monofilament falls slowly to the bottom. After a number of years

1960's: Deep-Water Triumph and Turmoil - 99

these non-dissolving strands of synthetic fiber create a web over a wreck which is as dangerous to divers as the sticky strands of spiders are to flies. The process of monofilament accumulation is accelerated when a trawler loses a net by accidentally dragging it into wreckage and losing it, for hooks snag nets far better than they snag wood or steel.

Add the fact that the Mud Hole is readily accessible to hundreds of head boats operating out of western Long Island and northern New Jersey, and there cooks a recipe for entrapment that would make the best chef weep. Into this dangerous graveyard of ships the avid wreck-diver plunged.

The most popular wreck in the Mud Hole was the Brazilian freighter *Ayuruoca*, sunk in 1945 after a collision with the Norwegian motor vessel *General Fleischer*. It lies at a depth of 170 feet. Because it sits upright on the bottom, the main deck rises to 140 feet and the top of the wheelhouse to 110. No one knew the identity of the wreck when divers first descended to its intact hull and rising superstructure. Because patches of oil occasionally rose to the surface, it was dubbed the Oil Wreck.

It took de Camp a fair amount of research to correlate what was known about the wreck with what was given in an obscure news account he found in the *New York Times*. After obtaining a museum photo of the *Ayuruoca*, he made an enlargement of the bridge and wheelhouse and took it with him under water. He swam slowly along the forward bulkhead and compared the wreck's features with those on the photographic print. The similarity between ship and wreck were striking enough to make the identification positive. Later, recovered chinaware with the name of the shipping line - Lloyd Brasileiro - confirmed the wreck's identity.

Other wrecks in the Mud Hole were deeper, approaching 200 feet, but they were visited less often because of the hazards imposed by the frightful conditions that were magnified by the depth. One of these was called the "Junior" and another lay just a few miles away from it. They were not identified until the 1970's (by this author). It should be noted that only the tall structures of these wrecks were explored at the time, not the lower hull or debris fields which lay deeper down.

UNKNOWN DANGERS OF DECOMPRESSION

Another wreck which lay at 170 feet but which was dived with less frequency was the "Virginia," now known to be the Norwegian freighter *Sommerstad*, torpedoed by a German U-boat in 1918. It lies sunk off the south shore of Long Island where the visibility often exceeds fifty feet. The wreck has been demolished to the point that very little recognizable structure remains, apparently the result of extensive salvage operations. It appears that the sides of the hull were blown out with explosives in order to recover the triple expansion reciprocating steam engine. The wreck is crisscrossed with fishing line due the considerable amount of angling for large cod and pollock known to frequent the site.

Evelyn Bartram was seriously bent on the *Sommerstad* after surfacing away from the anchor line: she became numb from the waist down and couldn't swim or walk. She was flown to a recompression chamber by Coast Guard helicopter. She responded poorly to treatment and was a long time convalescing, taking many months to recover. Even then she was left with permanent nerve damage in her right leg. This incident emphasized the risks of deep wreck-diving even if one avoided the hazards presented by the conditions found on the wreck.

NEW TALENT

Later, Bartram married the man who was perhaps the most skilled and most well-respected diver of the decade. His name was John Dudas. Dudas began diving in 1958 at the age of fourteen, when he lived in New Jersey. He eased into wreck-diving aboard the sailboat *Seal*, and later aboard the *Bottom Time*, Captain Charlie Stratton, out of Point Pleasant. Soon he was rubbing wetsuits with such divers as Ed Maleshewski, Ed Rush, Dick Hilsinger, and Hoffman and de Camp. He was part of the group that dived the *Pinta* and the *Stolt Dagali* when those ships first went down, and it was on the latter wreck that he gained prominence.

John Dudas' best friend was John Pletnik. They met after Pletnik came home from the Mediterranean, where he served with the U.S. Navy, and dived together until Pletnik went back to the Med as a civilian to dive for ancient amphoras. Dudas stood well over six feet tall; Pletnik barely topped five feet six. Because of their disparate proportions in relation to each other, they were called Big John and Little John, or sometimes Batman and Robin. Pletnik was not as interested in artifacts as he was in spearing fish. He could hold his breath long enough to shoot a cod at a depth of sixty feet, then reload and shoot a second one.

Dudas and Pletnik were important members of the team that answered de Camp's and duty's call when the *Stolt Dagali* went down. They displayed uncommon audacity in searching through the cabins and recovering for the ship's owners the personal effects of the crew: wallets, watches, change purses, and so on.

On older wrecks, Dudas had a knack for knowing where to look for artifacts, and an eye for spotting them among consolidated debris. He also did his research, and it was this that enabled him to locate certain items that other people only stumbled upon. He was a peaceful man who kept his own council and followed his own agenda: one which was not driven by any external imperative. He knew what artifacts he wanted, so he worked hard on obtaining them. When one was taken away by subterfuge, he was slow to anger. It was beneath him to consider anyone a threat to his talents or achievements. He was also among those who first dived such wrecks as the *Cherokee*, *Hvoslef*, *Jacob Jones*, *Moonstone*, *Northern Pacific*, and *Varanger*.

Dudas drifted through the scuba marketplace in a variety of jobs: an instructor, a representative for AMF Voit, and ultimately a dive shop owner with his wife, who started the business by making wetsuits in her garage. Throughout the years the quantity of unique artifacts that he recovered would make any diver envious. On the boat he lay quietly in a corner and read books, some on philosophy and Christianity.

Brass Fever - a New and Contagious Disease

Very early on in the annals of wreck-diving the salvage of maritime relics became a goal in itself, and the acquisition of these lost items became one of the driving forces behind the exploration of known sites and the search for undiscovered wrecks. These souvenirs from the past, these mementos of underwater adventures, these keepsakes adorning walls and mantelpieces, sometimes caused sad disillusionment rather than fond reminiscence - and occasionally strained or severed relations among active participants who passed themselves off to their fellow divers as friends.

The desire to recover or to otherwise possess a wreck's prime brass and bronze tokens created rivalries that reached beyond invective. Hard feelings arose between divers competing for goodies, and greedy individuals formed camps and chose followers to champion their side. Out of this animosity came secrecy and backstabbing. On occasion a diver found himself excluded from the mainstream because his talent posed a threat to those who were lesser endowed. Throughout the years, many a friendship has been lost over so common an item as a porthole.

Brass fever reigned as harshly as the lust for gold. If there was ever a ship called the *Sierra Madre*, it undoubtedly wrecked off the coast of New Jersey.

Definition of an Artifact

The reader should understand that these oft cherished remains are neither rare nor precious to the world at large. Bells, helm stands, and portholes can be purchased for reasonable sums from nautical antique shops a-plenty. They are trashed every day from ships that are broken up in scrap yards. And although such items are called "artifacts" in the wreck-divers' vernacular, they have no value as antiquities nor do they hold archaeological significance. Their worth is intrinsic only to those who covet them. On the open market they bring little or no return, and museums refuse to display them as meaningful objects of art or history.

A Bell for *Arundo*

A representative example of artifact avarice is the episode of the *Arundo*'s bell, which occurred in April 1966. The *Arundo* was a tanker that was torpedoed by a German U-boat during World War Two. The bell was initially found by George Hoffman when he was diving with either Dick Hilsinger or Ed Rush (the memories of surviving participants are vague). At that time the *Arundo* was unidentified, and it was hoped that the inscription engraved in bronze would establish the wreck's identity. Hoffman didn't have the tools required to remove the bell from its davit, which was pinned under steel beams and plates, so he had to leave it behind.

When the group next returned, the grapnel fell some distance away from the bell. This time Hoffman went down with de Camp, but they got disoriented in the murk on the bottom and spent a long time looking for obvious landmarks. Meanwhile, Dudas hit the water. Dudas possessed that same undefinable talent that enabled Daniel Boone to find his way through the wilderness without getting lost. He could distinguish referential features among the mass of encrusted wreckage, integrate the salient pieces into a recognizable pattern, and assemble an image in his mind like a completed cut-out puzzle. He swam directly to the bell and put a liftbag on it.

The liftbag did not have sufficient buoyancy to break the bell free from the wreckage. After he left, Hoffman and de Camp chanced upon the bell. Hoffman put a liftbag on Dudas's liftbag and sent it to the surface as a marker.

In those days, it was common practice to make the first dive with double tanks and the second dive on a single, the theory being that the limited bottom time of the repetitive dive precluded decompression. Thus, after a suitable surface interval, Dudas went down on a single tank with a hacksaw in hand. He sawed furiously on the iron davit, cut it nearly through, then, exerting his enormous strength, he stepped on the bar and broke it. Then he ran out of air and made a free ascent to the surface.

Hoffman and de Camp went in immediately afterward. Hoffman completed the job of breaking the iron bar, and sent up the bell on a liftbag. He and de Camp ascended right behind it. They surfaced in a thick fog with no boat in sight, and felt a few anxious moments till they glimpsed the *Sea Ranger* as they drifted past in the current. They kicked for all they were worth, Hoffman with the liftbag in tow, and soon were aboard with the prize. Then Hoffman claimed the bell as his own.

The Law of G. Magnus

Thus was created the "Law of G. Magnus," a phrase of Hoffman's invention which meant "he who floats it, owns it." (Hoffman claimed that Magnus was his middle name.) According to this law, it did not matter who found an artifact or how much time the finder spent trying to recover it. An interloper could dispossess him of it simply by sending it to the surface. For example, if a diver spent several dives removing all the bolts from a porthole but, when the last bolt came free, did not have enough air to fill a liftbag, someone lurking nearby could reap the rewards of his efforts.

The *Arundo* bell incident created great rivalries and set a bad example for the wreck-diving community. Like vultures, divers without the ability to find or remove artifacts lingered in the background and listened to the grapevine for news of unrecovered finds and projects in progress, then swooped in, often on a different boat, and carried off their ill-gotten spoils. This lowbrowed attitude engendered an era of contention and hatred, all in the cause of beating buddies out of artifacts.

Greed Never Dies

As a holdover from those primitive and ignoble times, there are those today who still apply this rule when it's to their benefit to do so. A recent example is that of the helm stand from the *Ocean Venture*, a freighter torpedoed off the Virginia coast by a German U-boat in World War Two. I sent the heavy brass piece to the surface in 1993. It sank before it could be retrieved, so I made another dive that afternoon, returned to the spot where I had sent it up, followed my safety line off the wreck to where the stand had come to rest, attached a second liftbag, filled them both, and refloated it. For some unknown reason, the helm stand sank again. That was the last dive of the day. During the coming weeks, in addition to spreading the word throughout the diving community that I wanted to return to the wreck in order to retrieve the stand, I continually signed on charters that were scheduled to go there. Every one was canceled because of weather.

Mike Boyle heard about my travails. Unable to find artifacts on his own, he took advantage of my situation and scooted out to the wreck on a charter on which I was intentionally not invited. He found where I had tied off my safety line, followed it straight to the stand, attached yet another liftbag, and succeeded in sending the stand to the surface and getting it to the boat. He then bragged about his achievement. To add insult to injury, he published an article in the Atlantis Rangers newsletter about how he pulled one over on me by beating me back to the wreck.

Avarice still runs rampant in the world.

UNWANTED RECOVERY

On a lighter note was a recovery from the *Ayuruoca* made by de Camp, Hoffman, and Walt Krumbeck. They found what they believed to be the captain's desk. It was a tall, rectangular object made entirely of wood and having drawers and pull-down shelves. Wrestling the piece out of the interior of the lower level of the wheelhouse was a monumental task, only slightly less difficult than rigging it with ropes and liftbags and sending it to the surface. As soon as Galluccio hauled it aboard it let off an unholy stench, driving him to throw it back into the sea. Consenting opinion held that the "desk" was in fact a water closet, contents included.

ARUNDO LEGACY

The saga of the *Arundo's* bell did not end with its recovery. For one thing, the name inscribed in bronze was *Petersfield*, a name not found on any known historical shipwreck list. This inspired Dudas to do research as a sideline. He soon found that the vessel that had originally been christened *Petersfield* later became the *Arundo*, and was traveling under that name when she was overtaken by the evil of German aggression.

Furthermore, Hoffman received so much grief from fellow divers over the incident - despite the fact that he was the original finder and Dudas was the interloper - that he agreed not to keep the bell in his home but to hang it on the *Sea Ranger*, the boat on which it was recovered. Later it was displayed publicly in a dive shop called Underwater Sports. This compromise helped to maintain a delicately balanced relationship between Hoffman and Dudas - for a while.

THE LURE OF THE ANDREA DORIA

The landmark year for deep wreck diving was 1966, when de Camp organized a trip to what has become known as the serious wreck-diver's mecca. When she first crossed the Atlantic in 1953 she was the largest and most luxurious liner in the Italian fleet, measuring some 700 feet in length. Her interior decoration was a paean to Italian Renaissance art: lavish murals reproduced from the works of famous masters graced the walls of the public rooms, while modern Italian artists added their touch by creating contemporary designs of startling contrast. Brass statuary, copper reliefs, and ceramic panels embellished the decor to create an atmosphere of elegance without falling into the trap of mawkishness. She was a ship to behold, a wreck to revere.

She was the *Andrea Doria*.

Eleven o'clock on the night of July 25, 1956 found the *Andrea Doria* westbound some two hundred miles east of New York City. Headed east from the Big Apple was the Swedish liner *Stockholm*. The two ships were on a parallel but reciprocal course which would have brought them to pass about fifty-five miles south of Nantucket. Due to a combination of events - fog and radar misinterpretation - the two mammoths came together in the vast reaches of the sea. The *Stockholm's* bow was torn off and stove in to a point seventy feet abaft the stem. The *Andrea Doria* was gored under the starboard bridgewing both above and below the waterline. Fifty-two people died at the point of contact. The *Stockholm* survived the ordeal and limped back to the City for extensive repairs. The *Andrea Doria* remained afloat for eleven hours: long enough for the survivors to abandon ship in lifeboats and to transfer to rescue vessels that responded to the call for help. Just after 10:00 a.m. on July 26, the *Andrea Doria* disappeared beneath the waves. She came to rest on her starboard side in 240 feet of water.

The *Andrea Doria* was first dived only one day after she sank. The forerunners who went down to the wreck (which was still burping air) were Peter Gimbel and Joseph Fox. Gimbel was an investment banker who was gradually quitting the business in order to follow his dream of underwater film making. Gimbel and Fox made but a single, brief dive: long enough for Gimbel to snap a few photos for a *Life* magazine exclusive. The beam of the ship was ninety feet, so when Gimbel and Fox descended to the upper side, they alit upon the shallowest part of the wreck.

Gimbel returned to the *Andrea Doria* two months later, again the following year, and for two weeks in 1966, this latter time with de Camp as his partner. He then conducted two major photographic and salvage expeditions, one in 1975, another in 1981. Despite these ventures, Gimbel was not a wreck-diver in the accepted meaning of the term: one who explores wrecks for the sake of exploring wrecks. Gimbel dived few other wrecks in his lifetime, an assertion which is not intended to diminish his underwater accomplishments, but to indicate the difference in his approach to the activity. He was also an infrequent diver, often letting two or three years go by between dives. De Camp remembers him as "gutsy and capable."

Two Innovations:
the Pony Bottle and the Decompression Reel

By this time the true wreck-divers were creatively adapting equipment and developing new techniques. De Camp and Hoffman originated the pony bottle and the decompression reel.

The pony bottle is a small tank which is secured to the primary tank(s) and fitted with its own regulator. It is used as an emergency air supply in the event a catastrophic failure befalls the primary tank or regulator, or if a diver runs out of air. Their first pony bottle was a Kidde fire extinguisher whose valve was replaced with a scuba valve. It was made possible only by the introduction of the single-hose regulator by Aquamaster. This is because a double-hose regulator may free-flow when the mouthpiece is not in the mouth, whereas a single-hose regulator does not.

The decompression reel was a broomstick wrapped with 1/4-inch sisal rope, which could be tied to the wreck and reeled off as the diver ascended from the bottom, allowing him to control his rate of ascent and to remain at the prescribed decompression stages by cinching off the line.

"If You Can Hook it, We Can Dive it."

A couple of months after the *Arundo* bell recovery, De Camp chartered the *Viking Starlite* out of Montauk, New York, Captain Paul Forsburg. The *Viking Starlite* was a 65-foot-long party boat that ran overnight fishing trips. The boat had room for all eleven divers and their mounds of equipment, but sleeping accommodations in the cabin were sparse, forcing the divers to repose on the deck wherever they could find the space. (Anglers generally fish all the time, so they don't require bunks.) This historic recreational trip ended nearly before it began because of a hurricane rushing up the coast.

The boat left the dock despite mountainous seas, and was more likely propelled by verve and audacity than by reason and common sense. With de Camp were some of wreck-diving's contemporary heavyweights: Jack Brown, Winston Chee, John Dudas, Dick Hilsinger, George Hoffman, Joe Holman, Robin Palmer, John Pletnik, Ed Rush, and Frank Scally. As the boat passed south of Block Island about three hours later, Forsburg called de Camp to the wheelhouse and showed him the state of the sea. Wrote de Camp, "We had to hold onto something every minute as the bridge swung violently from side to side."

Continuing to the *Andrea Doria* was out of the question, for that was another eight hours away in swells as long as the boat. Forsburg wanted to know if the divers were willing to chance the conditions to dive the wreck to which he had taken de Camp and dentist Walt Krumbeck the previous December, when the visibility had been so bad that they couldn't see what kind of wreck it was. De Camp made the classic gorilla diver statement that has come down through wreck-diving history in this form:

"If you can hook it, we can dive it."

He did, and they did. Fifteen minutes after Dudas and Pletnik hit the water a liftbag broached the surface. Then came an anxious delay for those waiting aboard while they completed a short decompression and came back with their report. Dudas was still floating in the water when he shouted, "It's a submarine." Suspended from the liftbag was the brass helm from the conning tower, which Dudas found lying loose next to the stand.

Submarine Discovery

Not until the following year did the sub reveal its secret identity, in a series of dives which is perhaps unique in the annals of wreck diving. Bill Hoodiman, an elevator constructor who worked with Hoffman, removed the handwheel from the hatch abaft the conning tower. In the company machine shop he machined a wrench to fit the stem of the bolt. With the extra leverage and assistance from Don Nitsch, a Long Island telephone company employee, he turned the hatch to the open position. Nothing happened until long after they left in grim disappointment.

A sudden explosion of air erupted on the surface with such violence that it shoved the *Viking Starlite* to one side. Wrote de Camp, "The sea boiled for a hundred feet around. . . . The air coming up had a very foul smell to it." For ten or fifteen minutes this phenomenon continued unabated.

De Camp and Hoffman were under water at the time. They saw the fountain of bubbles and thought that someone's regulator was free-flowing. But when they rushed to assist, they saw a massive V-shaped cone of air bursting from the hatch and expanding as it rose. The hatch swung wildly on its hinge as the compartment beneath it alternated between burping air and sucking in water to replace the lost volume.

The interior was a mass of silty turbulence for the remainder of the day. A few weeks later they returned to the site. Hoodiman and John Starace held lights at the hatch coaming as De Camp and Hoffman lowered themselves into the black interior. In a few short minutes they found a wooden box with a motor armature inside. Stenciled on one slat was "USS Bass." Subsequent research revealed that the *Bass* was sunk by aerial mines as part of an anti-submarine explosives test in 1945.

ANDREA DORIA OR BUST

But that was all in the future. After that first day on the *Bass*, the *Viking Starlite* pulled in to Block Island to wait for better weather. The boat left again the next morning in subsiding seas, and arrived over the *Andrea Doria* at 6:00 p.m. Forsburg established a bridle to hold the boat in place, then dropped a down-line with a grapnel at the end to snag the wreck. The divers wasted no time getting into the water for, despite the proximity of the summer solstice, the sun was low enough in the sky that most of its light reflected off the surface instead of penetrating to the depths. Add the stormy overcast, and it's easy to understand why their first glimpse of the wreck was dark and foreboding.

De Camp thought that he would never reach the lower end of the anchor line. "I'll never get back from this," he said to himself over and over. Not until he saw the glow of the hull taking shape in the formless void did he feel calmer. No one observed much of significance on that first dive, nor did anyone hazard a guess as to where on the wreck they alit, but just touching the hull was like sliding into home plate and making the winning run.

Forever after, all participants could declare with pride, "I dived the *Doria*."

Altogether they stayed five days over the wreck. Dives lasted between ten and fifteen minutes, with decompression adding another twenty or thirty minutes to each diver's time in the water. The superstructure was a monstrous incarnation of a block-long, multi-decked hotel. Windows grew innumerably along a seemingly unending hull. In places white paint was visible through the otherwise thick encrustation. The *Andrea Doria* was larger and more magnificent and spectacular than anyone ever imagined. Divers swam along the high side of the wreck like fleas on an elephant's back. Fleas who felt like gods.

The down-line tore loose the next day and had to be replaced. This time the grapnel snagged near the stern, offering divers an awesome view of the massive bronze propellers. The day after that, de Camp and Hoffman accomplished an incredible physical feat: they hauled the chain from the stern all the way forward to the bridge, traveling some 370 feet along the promenade deck at a depth of 170 feet.

This must have made Jack Brown happy, for it enabled him to find and recover a table-sized section of grated wood which was part of the platform that surrounded the compass stand on the exposed bridge. Dudas found part of a china cup - the good part, with "Italia" and the shipping line logo embossed in gold leaf. De Camp took photos to accompany a magazine article he wrote about the trip. He incorporated the pictures in his slide presentations.

Repeat Performance

The trip met with such enthusiastic response that de Camp organized a return engagement for 1967. Back aboard were Brown, Dudas, Hilsinger, Hoffman, and Rush; first timers included Evelyn Bartram, Jack Brewer, Walt Krumbeck, Cal Prater, underwater film maker Smokey Roberts (then at the beginning of a long and productive career), and Frank West. To increase creature comforts, the dive team arrived with hammers, nails, and lumber, and proceeded to knock together wooden bunks in which to sleep. The *Viking Starlite* left Montauk looking more like an immigrant ship than a party boat.

The weather was more cooperative this time. The most noteworthy event was Dudas' recovery of the binnacle cover and compass from the enclosed wheelhouse, from a depth of 200 feet. As an interesting sidelight, Italian film maker Bruno Valaiti dropped into the wheelhouse in 1968 with his camera. In the resultant feature film, *Fate of the Andrea Doria*, he shows a diver peering into the coverless binnacle while Valaiti narrates that the diver is peering into the face of the compass - the same compass that Dudas had removed the year before! Dramatic license, I suppose.

Scuba Comes into its Own

Apart from the obsessive wreck-diving zealots whose exploits I have portrayed, the 1960's saw the blossoming of recreational scuba nationwide. More and more people wanted to see for themselves the wonders of the underwater world. Scuba expanded beyond the stage of an aberration sought by extremists. It became an accepted divertissement for the masses. The tremendous increase in the number of divers turned the simple leisure time activity into big business. Equipment sales rose geometrically, producing an atmosphere that persuaded more manufacturers to enter the ripe market with newer and better products. Dive shops popped up everywhere.

No longer could someone toss a tank in your lap and say, "Don't hold your breath." Now there were people who charged money to introduce non-divers formally to the finer points of the activity: equipment familiarization, diving techniques, physiology, and safety. They called themselves diving instructors. Arguments arose among instructors as to whose training course was the best. After a while, instructors banded together against rival instructors and organized agencies which issued "official" certifications to those who successfully completed their standardized training program. Soon it became necessary to have a certification card in order to have tanks filled.

Dive boats proliferated toward the end of the decade. Some catered to divers rather than to anglers. Because tanks and gear bags took up so much space, not as many divers could fit comfortably on a boat as could anglers. Consequently, divers had to pay more money to meet the boat's stipulated charter fee. Dive clubs were formed by people who didn't necessarily want to dive every weekend, but who wanted space available for when they chose or had the opportunity to dive. That way, too, the financial responsibility was spread among all the members who paid dues to support the club's charter obligations.

ONE OF WRECK-DIVING'S DARKER MOMENTS

On the local scene, all presentiments of friendship between Hoffman and Dudas ended in November 1970. That was when Dudas and Bartram got married. For their honeymoon they went diving on the *Captain Chum*, which was chartered that day by Tom McIlwee with a covert purpose in mind: to catch the *Big Jim* on the *Moonstone*. Jim Dulinski was the only dive boat captain with the *Moonstone's* coordinates, a monopoly he maintained for those who faithfully chartered his boat. McIlwee and Robbins wanted the loran numbers for themselves, but Dulinski wouldn't give them up. When they heard that de Camp and Hoffman had chartered the *Big Jim* to dive the *Moonstone*, they carefully laid their plans to "jump" Jim Dulinski.

Chum Robbins hung back until he figured the *Big Jim* was grappled to the wreck and divers were in the water - making it impossible for Dulinski to move the boat - then he charged for the approximate location, spotted the boat in the distance, and homed in on it. Robbins circled the *Big Jim* to Dulinski's great annoyance. More upset than Dulinski, however, was Hoffman. When it dawned on him how he had been tricked he went into a tirade, shouting obscenities. The situation was not made better by the divers on the *Captain Chum*, who jeered at Hoffman's emphatic discomfiture.

Those, as they say, were the good old days.

NEW PROMISE

The 1960's ended with the formulation of a New York based club which catered primarily to the needs of dedicated and highly skilled deep-water wreck explorers. It originated as an organization for the elite - the Special Forces of diving - and not just anyone could join. Membership required the highest qualifications and was restricted to less than a score of individuals. This club exerted a tremendous influence on the future course and conduct of east coast wreck-diving, one that is still being felt today more than twenty years after its dissolution.

The organizer of the club was Elliot Subervi. The name he chose was Eastern Divers Association.

TEENAGE INTERLUDE

When the *Pinta* and the *Stolt Dagali* went down to a watery grave, I was still engaged in neighborhood exploration and adolescent high jinks. I was fierce on a bicycle, often riding at breakneck speed through the woods using trees as an obstacle course. I drove my bike off head-high drop-offs, careened down steep hillsides, made jumps over ramps and ditches, and plowed through water and mud. I rolled down hills in a barrel, and in the winter sledded down a giant hill on a refrigerator door. In both instances the object was to crash intentionally through the brush or over mounds of dirt. It was great fun.

In junior high I surpassed the 97-pound weakling stage, and by twelfth grade I was a 125-pound weakling. As a nerd I tended to get pushed around a lot by the bigger kids (which was nearly everyone, including the girls) but due to my inherent obstinacy I became an object of bewilderment to the male pubescent show-offs. Whenever someone shoved me (and I was easily shoveable) I stood my ground and stared straight into the eyes of my tormentor. I didn't push back or swing my fists, or respond with verbal abuse, but I never backed down.

The bullies didn't know how to treat a kid who was too small to beat up with any hope of earning respect from their peers, who was too stubborn (or too stupid) to surrender meekly to those of superior strength, but who resisted their self-appointed supremacy with what appeared to be silent daring. I didn't fit the pattern they had come to expect from those who were stationed lower in the pecking order as they perceived it.

Eventually, pushing me around became a game in which I was an active player. Sometimes they even landed light punches on my arms that were not intended to hurt, but to demonstrate that they could get away with it. Lest you think that I was overly bold for one my size, you should know that my actions belied my feelings: I was quite afraid of the outcome of every encounter. I didn't assert defiance out of courage or for fear of losing self respect (which I didn't recognize at the time), but because of the injustice of the situation. It was in my nature to resist, and I take no special pride in the fact - it was just something I was born with. I'd rather go down fighting than stand yielding.

The word "overachiever" didn't exist when I went to school, but if it had, it would not have applied to me. I was involved in no extracurricular activities. I didn't join clubs, I didn't play sports, I never went to a dance or a prom (I still don't know how to dance), and I wouldn't have attended the high-school graduation exercises if my mother had not insisted and bought a suit for me to wear for the occasion. I would have been content to let the school board mail my diploma to me.

CARS AND THE FRAT

A driver's license coupled with my parent's freedom with the car increased my mobility greatly, permitting me to develop a group of friends with whom I was considered an equal. We met in a garage which we converted to a fraternity house by laying down carpet and adding furniture that we obtained by weekly foraging expeditions known as trash-picking: driving slowly along the darkened streets on the night before trash collection, and picking up items still serviceable or repairable. We recycled long before recycling became fashionable.

The frat had no name, no bylaws, no committees, and no officers, just modest dues to pay the rent and a handful of members who liked to hang out and look for things to do. The most important member was Greg Carr because his mother owned the house of which the garage was a part. Other members were Jay Enright, Eddie Hackett, Tom Nash, Bill Reese, and Charlie Yearicks. Jay's younger brother Kenny hung out with us but was not an official member, and Kenny's classmate Bill Purtle sometimes came with us on our exploits. One or two others may have joined for short times, but if so I have forgotten their names.

Girls were forbidden to enter the frat house - not by us but by Mrs. Carr, who wanted to maintain a semblance of propriety in her garage. Sexual shenanigans had to be taken elsewhere. There were "parking" facilities at public places which were regularly patrolled by police. During one session of nighttime smooching my date suddenly shrieked in my ear when a snort impelled her to glance over my shoulder to see a horse's head leering at the window. Police patrolled the park on horseback so they could sneak up on cars whose antennas were swaying too wildly, hoping to catch couples in the act. And for this they got paid!

Our principal pastime was driving. Not that we had any specific destinations in mind. We just drove around northeast Philadelphia to see what was happening and, if nothing was going on, to make something happen. One of our more common capers was to tell ghost stories to the girls until they were sufficiently prepped, then take them to isolated cemeteries at night.

On one such occasion, on a dark overcast night in autumn, we found a fresh grave in which the dirt had sunk several inches and dried leaves had blown into the depression. I was becoming quite a prankster, so I shuffled irreverently across the leaf-filled excavation and crouched low to inspect the stone marker by the beam of my flashlight. Intoning deeply, I read the name and date of death, noting dramatically as I swung around that the person had died only a few months before. Kenny Enright and Bill Purtle knew that I was jesting, but told me afterward that the atmosphere I created was so realistic that even *they* felt shivers. Their girlfriends were terrified.

Feigning a look of horror, I stepped to the edge of the hole, groaned horribly, fell to my knees, then to my chest, flung my hands out in front of me, and dug my fingers in the ground at their feet. Then I squirmed backward into the depression by wriggling my legs and hips as I gouged deep furrows with my fingers. What I didn't realize was that my legs disappeared from view as they slid under the leaves so it appeared that my lower body was being swallowed up and that I was being dragged into the grave, yielding an effect far greater than the one I had extemporaneously arranged. Neither girl uttered a sound - their throats were paralyzed and they were too scared to scream. The next day, Kenny and Bill showed me where their girlfriends had gripped them by the arm. The surrounding, two-fisted purple bruises took weeks to go away.

My Own Wheels

After graduating high school I bought my own set of wheels. Seat belts were not stock items at that time, so when I drove the 1960 Ford Galaxie off the used car lot, I took it straight to an automotive shop and had seat belts installed. It seemed like a good idea. The car had high top-end speed but was not particularly quick off the line, although it could lay rubber with the best of them. If the 350-cubic-inch engine had had a four barrel carburetor instead of a two, it would have had more pickup.

I was captivated by speed, and now that the statute of limitations has expired, I can relate one of my high-speed escapades without fear of punishment. While the Philadelphia section of Interstate 95 was under construction, those of us in the frat who had cars would remove the barriers from the access ramps and drive on the unopened highway at night after the road crews had left for home. We flew along at more than 100 miles per hour unobstructed by traffic and meddling police (in patrol cars, not on horseback).

Right after the highway was opened to the public, Eddie Hackett rode shotgun for me to monitor the stopwatch we were using to test the accuracy of my speedometer at the high end of the scale. Instead of having the common circular dial, the Galaxie's speedometer was horizontal with numbers spaced evenly from 0 to 120. In the middle of a timed, two mile run (which at 120 miles per hour should have taken precisely one minute) we topped a rise and saw to our dismay a police car stopped on the right side of the road. The patrolman was walking toward a car he had just pulled over. We had a fleeting glimpse of a woman's face as she leaned inquiringly out her window.

Instinctively I took my foot off the accelerator. But I knew that I could not jam on the brakes at that speed without losing control of the vehicle or burning out the brake shoes. The pedal had to be pumped. Nor was there any way I could slow down to the speed limit (60 mph) in time to avoid being noticed. I punched the gas pedal instead.

The speedometer needle had already passed the last notch on the dial and was bouncing erratically, sometimes dropping out of sight below the faceplate. I switched off the headlights in order to darken the license plate. Eddie screamed *"What're you doing?"* but I paid him no mind. I knew I could stay in the middle of a three-lane highway without drifting over the shoulder. I must have outraced the radio waves. Three minutes later we were six miles away and shooting off the ramp without pursuit. With the lights back on we blended in with vehicles traveling innocently along the side streets. I dodged through several driveways until we were sure we were in the clear. After that episode we decided it was time to observe the posted speed limit.

The curious feature of high speed driving was the eerie sensation of silence attained after passing the 100 mph mark. The car seemed to float slightly above the road on a pillow of air, and the engine noise diminished dramatically as if muffled by some strange law of subsonic physics. All sensations within the car were suffused with a dreamlike hush and tranquility.

HIGH-SPEED GETAWAY

One of the biggest diversions in the 1960's was drag racing. None of us in the frat had cars fast enough to compete, but we loved to watch the hot rods and the cars with souped-up engines peel off the line in pairs, screeching and whining with power, getting rubber in second and third gears, and shooting across the finish line with thundering speed. There were no race tracks available to teenagers, and drag racing on the streets was forbidden, but a new industrial park erected adjacent to the Northeast Philadelphia Airport offered somewhat of a compromise if you were willing to bend the law. Two newly built streets were paved in smooth macadam as straight and as flat as a couple of runways, in order to access the parking lots of the commercial buildings situated so far from existing streets. In a sense these roads were very long driveways which were given county street status. They were thoughtfully provided with lights mounted on the tops of tall poles.

The two roads formed an L, the short leg of which was half a mile long while the long leg was more than a mile. The long leg was perfect for racing and soon became legendary. It was called Decatur Road.

During the day the roads were used by employees who worked in the buildings which the roads were built to service, but at night they were deserted. Most older teenagers in Philadelphia were aware of the nightly performances that took place there. On an average night as many as fifty cars lined the curbs, with several hundred teenagers (and some people in their early twenties) waiting for events and having good clean fun. By common consent no alcohol was permitted (we didn't know about drugs). This was so the police couldn't come down on us for underage drinking. (The drinking age in Pennsylvania was eighteen.)

The races were supervised with professional conduct: there were flag men, timekeepers, and spotters at the end of each road where they connected to the city streets. The spotters flashed their headlights as a warning against the approach of the rare motorist who knew about the roads and used them as a shortcut to avoid traffic on the more regularly traveled streets. Activities ceased and way was made for such motorists, who were waved through courteously and without intimidation.

Enforcers were designated to maintain crowd control. For safety reasons, as a match was about to begin, the enforcers walked along the two lines of cars to ensure that no one was standing in the street. According to the rules we had to be in the car, on the car, behind the car, or between cars: places where one couldn't be struck by a dragster gone astray (although that never happened). For better viewing, most of us sat on the roof or on the hood or trunk.

The starting line was painted close to the junction of the two legs of the L. The finish line was carefully measured and painted a quarter mile away. This gave the race car drivers plenty of empty roadway for decelerating their cars.

The police eventually learned about the drag racing on Decatur Road. They couldn't condone the activity, but most of the time they left us alone and the races went on without official protest. Every once in a while a lone cop in a car would break up the races by stopping in the middle of the street in front of the starting line. He would get out and tell everyone to go home. We groaned, but accepted the fact that the races were over for the night. In actuality, by the time he got there many of the cars were already pulling out, since we had been warned by a headlight signal from one of the two spotters who were innocently "parking" with their girlfriends at the end of each road. There were never any confrontations or protests, or even harsh language. It was all part of the game.

After a while the police took a tougher position that severed our unspoken understanding. They could never observe a race in progress or catch contestants in the act because of the efficient remote warning system. Instead, if they got there providentially when two cars were idling on the starting line, they began a policy of ticketing the drivers on the presumption that they were about to break the law. More often than not, the police returned to the doughnut shops in frustration, having made a raid at a time when no races were proceeding.

116 - The *Lusitania* Controversies

Harassing teenagers eventually became a favorite police pastime throughout the city. Immediately after midnight, the cops pulled over every car that was driven by one who looked adolescent, hoping to catch a driver with a junior license. They also demanded identification from the passengers who might be minors breaking curfew. If that failed to produce a violation, they looked in the trunk and inspected the car to see if they could issue a ticket for a burned out brake light or a broken turn signal. (An adult was issued a warning for such minor offenses; an adolescent received a ticket in his own name even if the car belonged to his parents.)

I was stopped so many times for "car checks" that I kept a list of the patrol cars' numbers. Once, when I couldn't see the numerals painted on the door, I ran back with pencil and tablet in hand. The cop asked what I was doing. I wrote down the number and showed him the book, and explained that I needed only one more number to have been stopped by every car in that district. After a while I sought out the patrol cars in neighboring districts in order to flesh out my list. All I had to do was to cruise past a patrol car after midnight. I was sure to get pulled over. The cops had nothing better to do, and neither did we.

I was once given a ticket for stopping in a shopping mall parking lot after the stores had closed. The charge was trespassing. Since I was attending college downtown at the time, it was convenient for me to go to traffic court and protest. About twenty-five people were in line ahead of me, all adults fighting bona fide violations. After listening to their stories, the judge found every one of them "not guilty." I was the only teenager in the group. I had to pay the fine plus court costs. I hung around for a while and noted that every adult after me was also found not guilty. So the policy of harassment went deeper into the establishment than the flatfeet on the force. Yet we were the very people who, when those in authority wanted their homes and families protected from foreign aggression, were chosen to die for their country.

It seems to me that the alienation that exists between youth and authority is less a condition of youthful rebellion than a condition of authoritative control. The desire for domination which is displayed by certain individuals - particularly parents, politicians, and bureaucrats - eventually pervades the society of which they are an influential part. If the establishment did not place barriers before alternative points of view and new modes of thinking, contention would neither be created nor proliferated. History has repeatedly demonstrated that any person (or people) who is coerced against his will, opposes such coercion and will respond with a degree of resistance which is perceived as a threat by those who believe that they have a right to direct the fates of others. The inevitable result is disobedience on an individual level, insurrection on a larger scale.

1960's: Deep-Water Triumph and Turmoil

In this wise the Philadelphia police and judicial system conspired to "get" the teenagers who had the temerity to determine the volition of their own lives. The city, in its infinite wisdom, passed a law making it illegal to *watch* drag races. Now when the police arrived they would ticket not only the drivers of the two cars on the starting line, but could detain everyone in sight. And this they proceeded to do.

Several mass arrests occurred in which hundreds of teenagers were booked in one fell swoop because they chose to gather at Decatur Road instead of at "approved" locations such as movie theaters, bowling alleys, or fast food restaurants - where, by the way, they were expected to put money into society's till. These "drag busts" occurred even when no drag racing was being conducted. That Decatur Road had become a "hang out" favored by teenagers was ignored by police, who had initiated a prejudicial plan of guilt by proximity.

I barely escaped one of these adolescent pogroms. One night a fleet of patrol cars closed in at high speed from both directions at once and barricaded the road with their cars. I leaped behind the wheel of my car, cranked up the engine, and took off so fast that I didn't even know if the rest of guys were with me. There was a screaming, squirming mass of bodies in the back seat, with arms and legs entangled like a bucket of earthworms, and someone's legs dangling out of the front window. I was still pealing rubber when a policeman blew his whistle and yelled, "Hey you, stop!"

Not likely. I had a glimpse of a startled face and expanding eyes as the cop spit out his whistle and jumped back out of the way before I shined his shoes with rubber. Other cops were discharging from a car behind him. I was going fifty miles an hour and approaching the bend fast when I twisted the wheel hard and threw the car into a controlled skid. With my headlights spearing the rearmost patrol car it must have looked to them like I was going to ram. One policeman rolled back inside, the others scattered. But I knew what I was doing.

The law of inertia applies equally to society's resistance to change as it does to automobiles traveling at a high rate of speed. Anyone who has ever lost control of a car on the ice knows what I mean: turn the wheel and touch the brake and the car goes into a spin. The orientation of the vehicle changes but not the direction of travel - because ice is nearly frictionless. Every northerner learns that a vehicle can be coaxed out of a spin before control is completely lost by releasing the brake and applying a little gas. On a surface that provides sufficient friction for the tires to get a grip, a skid can be induced and controlled in a similar fashion. The trick is to apply enough gas to feed the slippage while maintaining forward momentum, then easing off the gas in order to permit the screeching rear tires to gain just the right amount of traction needed to propel the car in the direction in which it is facing. Proficiency takes a little practice. I had plenty - and besides, I was a natural.

In this manner I skidded with precision around the ninety-degree bend while seemingly pivoting around the blockading patrol car. Only three cars escaped apprehension that night. Several hundred teenagers spent the night in jail until they were bailed out by their parents the next morning. I have yet to understand why government seeks to intrude itself upon the lives of its constituents when those constituents are minding their own business and are not involved in any acts which could be considered harmful to others. Passing a law against an activity does not make that activity bad, wrong, or immoral. By definition, it only makes it illegal.

Abandoned Houses

Another way we found excitement was by exploring condemned or abandoned houses at night - and the darker the better. Technically this could be considered trespassing or entering since the property had to be registered in someone's name, but the houses we crept through were disintegrating through neglect because there were no resident owners, and those that we explored were far beyond repair. Eventually, our favorite "haunted" houses either collapsed on their own or were razed for new construction.

We sought out these forsaken dwellings wherever we could find them by driving around the countryside in the daylight hours and by asking friends and relatives if they knew of any broken-down ruins about to be demolished. Many of these were isolated farm houses which were large with many interconnecting rooms, and often had multiple stairwells. Usually we conducted our excursions unobserved and left with no one the wiser. Sometimes we played tag with police who did not take kindly to our harmless high jinks. This added spice to the adventure.

One time patrol cars surrounded the building while two members of the constabulary entered the front and back doors. We escaped to the roof and climbed down a drain pipe, then slunk off into the woods undetected. Another time a solitary patrolman seemed intimidated by the presence of six healthy teenagers. In order to put him at ease I walked up to him calmly and explained exactly what we were doing. He seemed relieved, if nonplused. He ordered us off the property without even asking for identification. A couple of times we were caught and taken to the local police station, only to be released after questioning. They couldn't understand why anyone would want to tiptoe across a musty attic inhabited by pigeons, or slosh through a damp, moldy basement that was crawling with spiders and rats.

It was a cheap thrill that did no one harm.

The Call of Adventure

The exploit that I recall the most fondly was my last before the complicated commencement of adulthood: a climb to the top of Bowman's Tower on an overcast, rainy night. The ancient stone lookout tower rises 150 feet above the top of a tall hill overlooking the Delaware River just upstream from Washington's Crossing, where the reluctant revolutionary war general and later first President made his famous midnight Christmas Eve raid against British troops encamped on the opposite shore. During the day, Washington's Crossing State Park was a favorite picnic area. On summer weekends the grassy grounds was visited by hundreds of tired city folk seeking surcease from the pressures of civilization, and children eagerly looking forward to a day of play in the sun: At dusk the parking lot was chained off and entry to the "public" facility was not permitted till dawn: another inanity of modern ordinance.

Bill Reese parked his black '57 Chevy in a pull-off at the base of the hill. Along for this excursion were Jay Enright, Kenny Enright, Bill Purtle, and Charlie Yearicks. I explained how to breach the park's defenses and how to climb to the base of the tower up the narrow, winding road that was closed to nighttime traffic. We were undeterred by the imposing darkness and the cool April showers portending.

We crouched in darkness until the country road was clear of passing vehicles, then made a dash across the black macadam and into the woods beyond. We bushwhacked just inside the tree line, out of sight from the road. By skirting the picnic grounds we intersected a dirt path that arched across the lower approaches to the main rise of the hill, then reached the deserted access road before it began its steep ascent. About this time it started to drizzle, but no one wanted to turn back.

A slight mist hovered close above the day-warmed ground to mingle with the cool falling rain. The result was a ghostly shimmer which caused the shadows and the silhouettes of trees and shrubs to sway, creating a montage of evanescent shapes, existing for a moment but gone upon closer examination. By letting my imagination wander, I could envision a forest full of stalking mythical creatures from beyond the dimensions of reality. Of course they were only apparitions of the mind, and I knew it.

Our light-hearted banter grew louder the farther we got away from any chance of being heard. The trick of exploring forbidden territory is in knowing that only the perimeters are guarded. Once inside, one has free reign from those who think their fortresses are impregnable.

As the rain fell harder and our clothes soaked through, we laughed about how ludicrous our venture now seemed under the circumstances, although none but Yearicks expressed a desire to call it quits. Even so, I knew that for me the rain and the lawful prohibition against our exploit contributed to making the experience notable and worthwhile. Walking up the road on a sunlit afternoon would have been meaningless: a trifling memory soon forgotten.

After due course we reached the top of the hill. In the broad clearing that dominated the flattened summit, Bowman's Tower stood like a mighty monarch delineated against the blackened sky. Standing at the base of the circular stone stronghold and yanking on the handle of the massive wooden door, which was locked, I felt as if I were demanding entry into the castle of a fabled ogre in medieval England. By this time we were thoroughly drenched. Jay Enright, Reese, and Yearicks had had enough kicks for one night. They wanted to return to the car to get warm. I had more activity planned and didn't want to leave. Kenny Enright and Bill Purtle elected to remain. I tried to convince the others to stay, but they were resolute. Resignedly, I gave the three departees directions for the short way down: a shoulder-wide dirt path that dropped steeply through the trees toward where the car was parked. At the fork about midway down I cautioned them to bear left. They promised to wait for us in the car, but asked us not to be too long.

The tower was about twenty feet across at the base, and sloped inward slightly as it rose. It was constructed of thick rectangular stones and had a castellated top very much like the rook on a chessboard. Paneless windows coiled upward to coincide with the spiral staircase within. The lowest portal lay fifteen or twenty feet above the ground. The cement between the courses of stone was grooved barely enough for me to gain purchase with my fingertips. The stone was wet and slippery. Clinging to the wall like a fly, I climbed slowly upward by pulling with my fingers and pushing with the slight ridge of my shoe soles.

Scaling stone walls was child's play for me - literally. My house had a chimney made of blond Tennessee stone, and I had been climbing to the roof for years.

I gained the height of the window and got a good grip on the sill. The black void inside was barred by an iron grate which consisted of two vertical and two horizontal rods, forming a crosshatch whose central opening was about a foot wide and a couple of inches more in height. I grabbed one of the rods. Now there was no way I could fall. By leaning back against my arms, I got my feet onto the sill and snaked them through the grate's central opening. I turned on my side to squeeze my hips through. I wriggled backward till my shoulders wedged me tight against the rectangle of rods. For a moment I was stuck in that position, with my legs and feet swinging in mid air inside the tower. By angling my shoulders so that one went in first and then the other, I forced myself completely through the opening until my body was swallowed up by the intense blackness of the interior.

My feet touched nothing when I lowered my body to the full extent of my arms. My grip on the rods held the full weight of my body. I remained suspended in that position for a long time. I could see nothing but blackness. Not a single photon penetrated to these nether reaches. I felt an eerie sensation of dangling over a black pit of infinite depth. Yet, I knew from previous daytime visits that a staircase wound around the inside perimeter of the tower. It must - *must* - lie just below my extended toes.

I steeled myself for the worst, took a very deep breath, and with a leap of faith let go. I fell about two inches. My heart was beating faster than normal despite the evident relief.

I climbed back up to the portal and crawled halfway through. I shouted down to Kenny and Bill that the way was open. First one and then the other scaled the bulwark. I showed them how to squeeze through the iron grate and eased them down to the steps. For their own sense of security, I helped them walk inward with arms outstretched until they felt the inner iron railing. Then we began our ascent.

With one hand dragging against the cold stone wall I followed the stairs around the inner curve of the tower. Kenny and Bill Purtle were right behind me. We could see nothing but faint patches of light from the upper windows as we passed them by. Our voices echoed hollowly in the cavernous interior. Around and around we went, climbing higher, until about a hundred and fifty steps later we emerged onto the lower observation deck. This was a square room whose roof was supported by a thick stone column in each of the four corners. We dashed about from side to side to enjoy the view. Because the rain had let up, we decided to go up onto the roof deck.

An enclosed spiral staircase only slightly more than shoulder wide wound upwards, much like a hidden corridor in an ancient English citadel. Two people could not pass in it, and on crowded weekends someone was always having to back up or call ahead. We felt our way up in the dark. The top deck was surrounded by a thick stone bastion. From here we had an unobstructed view of the countryside in all directions.

The Delaware River was a thin silvery ribbon that was bordered on both sides by trees. The lights of houses dotted a surreal landscape which in the dark and overcast was shapeless. The roads were invisible, but headlights and taillights moving with painful slowness marked the routes through the forest and farmland. Although I could not see it, I pointed out where the car was parked, then indicated the lay of the land as I had seen it during previous daytime visits.

At the base of the hill I noticed something that seemed to be out of place. Twin white cones from a pair of headlights made sweeping curves through an area that I thought lay within the boundaries of the park. The headlights came to a halt and a car door opened, the action revealed by the weak emanation of a dome light. Then a hand-held flashlight moved dimly away. I was certain that someone was searching the grounds, where access was prevented by firmly locked chains.

Suddenly it came to me: only the police had access to the area. And if the police were searching the parking lot they were probably looking for us. We watched the fruitless investigation for several minutes, laughing all the while, before realizing that something must be dreadfully awry, and that those waiting for us in the car must have somehow aroused suspicion. That meant that the secrecy of our activity had been compromised. Then we spotted more headlights sweeping other areas of the park, and knew for sure that a full-scale investigation was underway. This boded ill.

We retreated into the tower and descended the invisible spiral staircase. The way down was more alarming psychologically than the upward climb had been, for it was like a dark descent into the very pit of Hell. Every step might have been a plunge into eternity. Only intellectual acceptance of unperceived reality furnished the resolution to continue. I did not feel for the next lower step with my foot. I just knew that it had to be there.

I stumbled momentarily when I reached the solid floor, for I was in the act of reaching down to where I expected the next lower step to be. After a moment of disorientation, I warned Kenny and Bill. I explored the circular base of the tower with my hands, felt the inside of the locked door which was chained from the outside, then backed up the stairs to the lowest window. We had passed it in the dark as there was no way to know that it was the bottommost portal. We squeezed through the grate and climbed down the outer wall to the ground. I led the way down the dirt path, almost missing the left fork at the split about halfway down the hill. Near the roadbed I veered off the path and struck out through the woods in order to emerge from the trees adjacent to the car so we could assess the situation. We came out right on target, but a deeply cut stream blocked easy passage. We slid down the steep, ten-foot-high embankment, danced across the rocky stream bed, then scrambled up the other side and crawled through the brush to the edge of the road. We peered through the sparse foliage cautiously.

A patrol car with the dome light on was parked behind the Chevy, so we knew that the jig was up. But we didn't know how much of our activity had been revealed. Rather than pounce onto the road from the trees, we backed up, recrossed the stream, and walked through the woods for a quarter of a mile, then gained the unlighted country road from around a bend and walked openly along the road in the rain. With this approach we could say that we had not been inside the park

1960's: Deep-Water Triumph and Turmoil - 123

at all but had only been walking around it. These precautions came to naught because Reese and Jay Enright had already blabbed. And they hadn't even been tortured.

I walked boldly up to the driver's side of the patrol car and bent down by the open window. "Hi. Are you looking for me?"

The policeman looked up from his clipboard, held my eyes and smiling face for a moment, then stared down at the list of names and addresses he had in his lap. "You must be Gary Gentile."

I glanced at Bill Reese, who was sitting on the front seat by the other door. He raised his eyebrows resignedly. "That's right."

My two companions passed by without a sideways glance and climbed into the Chevy. The policeman said, "And they must be Kenny Enright and Bill Purtle."

"Right again."

Then came the questioning period. I showed him identification and verified names and addresses, even gave him my phone number. I still didn't know what story he'd been told, so I let him take the lead about what we were doing and where we had gone. As it developed, he already knew that we had walked up the back road to the top of Bowman's Hill, so there was no sense denying it. But I played down what we did on top, saying only that we wandered around a bit after the others decided to come down.

With an assumption that was either perceptive or coincidental, he said, "You didn't go in the tower, did you?"

"You can't *get* in the tower. The doors are locked at night with a chain."

He thought about that for a moment, then said, "All right, I'm gonna let you go. I know where to find you if I want you. And *he's* responsible for driving you here." He jerked his head toward Reese. "But I'm going up there in the morning to check it out, and if I find any damage you'll all be in trouble. Understand?"

I understood that we had nothing to worry about. Throughout all our unconventional activities we never committed harmful deeds to people or their property. We didn't deface tombstones in the desolate cemeteries we visited, we didn't smash windows or break up furniture or ruin the structure of the "haunted" houses we entered, we didn't write graffiti on the walls of the sewers we explored - even though the likelihood of anyone seeing a painted disfigurement was doubtful. Under the conditions laid down by the policeman that night, we were guilty of nothing more than a stroll in the park - quite literally - after closing time.

We were released under our own recognizance. Reese let himself out of the hot seat and headed for his car. I stepped away from the window and started to follow, but the patrolman called me back. He looked up at me with a half-hearted scowl that barely disguised his chagrin. With a slow shake of the head, he said "What the *hell* are you kids doing out on a night like this, walking around in the rain?"

I grinned expansively. "Just having fun." And I knew that despite his apparent annoyance, we had interrupted a night of boredom and had given him a story to tell the rest of the men at the station, about the crazy kids from Philadelphia who had climbed Bowman's Hill on a dark rainy night instead of watching sitcoms on television.

During the forty-five minute drive to Philadelphia, amid laughter and yelling and wild accusations, I heard the other side of the story. Yearicks couldn't keep up with Jay Enright and Reese. He missed the fork in the dark and took the wrong path to the right. This brought him out on the other side of the hill, where nothing looked familiar and where he became hopelessly lost. After wandering aimlessly along another road and becoming more confused, he knocked on the *back* door of an isolated house and asked for directions from the woman who lived there alone. She tried to tell him where he was, but this didn't help because Yearicks didn't have a clue as to where he was supposed to be. He didn't know that we had parked on the road by the river. He left and wandered around some more. A few minutes later he returned to the house and again knocked on the back door - back instead of front because that door opened into the kitchen, where the woman sat at the table, and no other lights were on in the house. He reiterated his plight. By this time the elderly woman was getting nervous. She got rid of him, then called the police and complained of a prowler in the neighborhood. She had every right to be scared. Yearicks was well over six feet tall, blonde, broad chested, and brawny. His gentle and easy-going manner went unnoticed under the circumstances.

Patrol cars were dispatched to cruise the neighborhood in search of suspects who might be casing homes to burgle. A patrol car passed by Reese's Chevy parked along the river road and turned back to investigate. By that time Reese and Jay Enright were sitting idly in the front seat waiting for the rest of us to return, and wondering what had happened to Yearicks. When the patrol car passed, they ducked out of sight. And when they heard the car stop, turn, and pull in behind them, they crawled down low under the dashboard and pulled a blanket over them, so a casual glance into the car would reveal nothing but darkness. It was in this compromising position that the patrolman's bright flashlight picked them out. They were understandably embarrassed.

During questioning, they learned about the "big blond guy" who was scaring little old ladies out of their wits. Radio calls kept the county police force apprised of current events, and of the apprehension of two possible accomplices. Reese and Enright overheard the conversations. With such serious accusations being made, they quickly came clean with the sordid truth, which the patrolman found barely believable. The search for suspects was then concentrated on the park, and this is what we saw occurring from the top of Bowman's Tower.

Reese and Jay Enright were sitting in the patrol car when Yearicks walked by without so much as a word or a sideward glance and plumped into the back seat of the Chevy. Reese said, "That's Charlie Yearicks."

By the time I arrived with Kenny Enright and Purtle, there was nothing left to say that had not already been admitted - except for the more illegitimate feat of scaling the outer wall of Bowman's Tower and penetrating the inky interior. That we did not mention until we were well out of the county.

At the end of all the banter about retributions threatened by the police, Reese said scoffingly, "So what can they do to you? Send you to Vietnam?"

They could. And they did. My journey began about three weeks later.

Descent into Horror

The hueys flew in low over the jungle that bordered our tight perimeter, barely cropping the trees in their endeavor to avoid enemy fire. Snipers had picked us apart during the night, and we were anxious to get out of there even if our next assignment was a search and destroy mission against a suspected Vietcong ville.

My belly was full of cold C-rations: ham and eggs chopped, because I was the only one in the company, perhaps in the whole army, who liked them. Clipped to the front of my web gear were six fragmentation grenades. The ammo pouches on my belt held one-hundred-sixty 5.56-millimeter rounds for the M-16 that I held loosely by the rear sight. A pair of canteens, rain gear, sundries, extra socks, and "C's", completed my accouterments. Raiment was a pair of dirty jungle fatigues, well-worn boots, and a steel pot on my head. We traveled light, depending upon resupply if we got into a major firefight.

The distinctive whup-whup-whup of rotating blades sent a chill along my spine, for I equated the raucous resonance with incipient fear and danger. More than a dozen gunships touched down in the pre-dawn dust of the dry rice paddy where nearly a hundred armed men crouched waiting. As soon as the landing skids touched the dirt, we ran hunched over for the doorless and seatless cargo compartment. I was the last one to board. There was no more room inside, so I had to sit on the outer edge with my legs dangling over the side of the platform, next to the door gunner who glanced at me over the barrel of his machine gun. No one grinned or smiled.

We took off instantly - if in fact the huey had ever landed. Usually they maintained a hover with very little weight of the machine on the ground. The operation was more like a touch-and-go than an actual landing and takeoff. We zoomed away and gained altitude simultaneously.

From my position half-out of the cargo compartment I had a perfect view of the lush jungle below. Vietnam was a beautiful country when seen from the air: a patchwork quilt of wet and dry paddies whose symmetry was artistically rendered by alternating hues of green and brown. Great curved swaths of untamed jungle separated concentrations of paddies as if a painter's brush had made broad verdant strokes across a living canvas. Randomly placed mounts like giant gumdrops rose hundreds of feet almost vertically from the rice paddy plains. In the mountainous regions called the highlands, undulating triple-canopy jungle presented a picture of living radiance that merely hinted at the profusion of lifeforms hidden beneath the towering tree tops.

Only from the level of the ground did the illusion of innocent splendor disappear. The dry rice paddies were barren and lifeless, the wet ones were covered with scum. The jungle was often impenetrable by man and lurking with tigers and boars and snakes and monstrous poisonous insects. Disease and squalor inhabited the hamlets, poverty was the watchword. And worst of all spread across the land was the ugliness of war.

The flight of hueys swung north along the coast in the suffused early morning light. From an altitude of five thousand feet I watched tumultuous waves in the turquoise blue water of the South China Sea crash against the white sandy beach. All below appeared calm.

We were right over the surf line - albeit a mile high - when the huey banked sharply for a turn. So pronounced was the list of the aircraft that the deck I was perched on tilted instantly to the vertical. I gulped. My left hand gripped the bulkhead fiercely, my right hand squeezed my rifle. No longer was I looking out over placid pleasant scenery. Now I stared straight down into the jaws of death at the end of an accelerating drop, with no restraints against my body to prevent me from falling out. I teetered over the fathomless abyss, unable to do anything but hold on tight. Inertia pinned me to the metal deck with the same principle of physics that prevents water from spilling out of a bucket that is swung horizontally in a circle. The huey leveled out, and none but I knew of my supreme moment of terror.

The flight spiraled downward at a more moderate pace that was timed to reach our destination at sunrise. The hueys gathered into formation like a flock of Canada geese. We dropped down close to the surface of the sea, then flew toward land out of the rising sun - just like the scene in *Apocalypse Now* except that we had no speakers playing Wagner. When the beachside village hove into view the hueys pealed off to the left and right and announced our presence with machine-gun fire. We swooped up to get a better view of the so-called enemy village and to allow the door gunners to sight targets trying to escape. We encountered no resistance, so the gunners took out their frustrations on the water buffalo wallowing helter-skelter in the paddies.

We circled the landing zone once, then a voice behind me shouted over the noise of the screaming engine, "Get ready!"

There was no time to recover from my earlier fright - a greater one was coming. I swallowed my heart as I slid off the platform and planted my boots on the skid; I steadied myself with one hand. We were still several hundred feet in the air and racing forward at nearly a hundred miles per hour. The wind whipped tears from my eyes. The village was ensconsed within patches of jungle and stands of thick bamboo, all of which was surrounded by fields and paddies in which grew the simple staples on which the peasants managed to survive.

The huey angled down over a dry paddy. The downdraft from the rotor blades churned the loose dirt into a turbulent cloud of dust that temporarily blinded me. Then our forward motion ceased and we plummeted sickeningly toward the earth.

"*Go!*"

Quick deployment and a rapid dispersal were necessary lest the huey be shot down by enemy guns or rockets, or lest small arms fire decimate the troops. When I looked past my boots I saw nothing but a beige, swirling miasma with no sign of the ground. We could be ten feet in the air - or a hundred. Above the roar of the engine I heard automatic weapons fire and the whump of mortars landing nearby.

A hand smacked hard against my shoulder, and shoved. I fell overboard. I braced my legs for a drop of indeterminate height. When I hit the ground after falling but a foot, my stiffened muscles could not absorb the shock. I tumbled onto my knees. Only my outthrust hand prevented me from crashing flat onto my face. An instant later a pair of boots scraped my side and someone stumbled over me. I scrambled forward on all threes - I held my rifle ready to fire - to avoid being trampled by the squad members behind me. Then I struggled to my feet and advanced in a low crouch.

Bullets whizzed by from all directions. I did not return fire because I saw nothing but airborne dust and dirt, now made worse by the departure of the hueys. A stark feeling of abandonment welled up inside me when the firepower of the gunships no longer backed us up. Now we were on our own.

Corporal Yawn indicated a direction and yelled in my ear, "Keep moving."

I did what my squad leader ordered. I glanced around and saw only two or three men through the tan haze. As I got farther away from the turbulance, the visibility in front of me increased.

Conducting reconnaisance by fire, I shot a few rounds into the tree line that demarcated the side of the paddy bordering the ville. I crouched lower and scurried quicker, thankful when I saw a dike that rose about fifteen inches above the dry paddy bottom. I hit the ground ten feet before I reached the dike, then low-crawled frantically to the meager protection provided by the rampart.

The cacophany created by bursting grenades and the steady but sporadic rifle and machine-gun fire produced an atmosphere of frenzy. I poked my head above the dike, but only high enough to enable me to see ahead. A dense tangle of brush protected a bamboo thicket that acted as a natural rampart for the village. It had been cultivated for just that purpose. I could see nothing beyond - and certainly no targets at which to launch an offense.

Seeking guidance in what course of action to take, I looked left and right - then left and right again. Then over my shoulder. Not a soul was in sight.

I was alone.

I had advanced so fast and so far that I had outdistanced the entire company. Instinctively I wanted to jump up and dash back to the perception of safety in the numbers of my outift. But then I envisioned what my mates might behold: a soldier looming out of the haze, olive drab in color, and running toward them bearing a rifle. Would they mistake me for an enemy soldier?

My heart thumped wildly. What was I to do to avoid being gunned down by friendly fire? The dust began to settle. Shapes coalesced. I turned away from the oncoming troops, placed my rifle on top of the dike, and fired into the bamboo thicket - not *at* anything, but in the direction which would illustrate whose side I was on.

Yawn stooped down beside me. "Good work, Gentile. Keep up fire superiority while we make a break in the bamboo." If he only knew.

I changed magazines and resumed firing. The rest of the squad plumped down on either side of me, as did the men from the other platoons forming our single rank. Several men hurled grenades high enough to clear the tops of the stalks, so the grenades did not bounce back at us. Machine guns came on line and shredded the dense foliage into confetti. Then we chopped our way through the splintered barrier and into the outskirts of the village. We stayed off the paths because they might be booby-trapped.

By the time we reached the deserted hooches the enemy had broken off the engagement and had melted into the jungle on the far side of the village. We found no bodies, discovered no abandoned weapons, and saw not a single enemy soldier in the flesh - either dead or alive. The phantom army of Vietcong guerrillas fought on its own terms much the same as the ragtag Continental soldiers of the American Revolution played cat and mouse with British redcoats. History was

1960's: Deep-Water Triumph and Turmoil - 129

repeating itself with the tables turned and with previous lessons forgotten by U.S. Army tacticians.

At day's end, when the tunnel vision of combat had worn off, I wondered how much of what I had seen and thought and felt that day was real, and how much was imaginary. Thirty years later I still don't know.

Caught in the Draft

I was a hopeful college student attending Temple University when the overseas "conflict" began. Like most people at the time, I had never heard of Vietnam. And now, like most people today, I can never forget it.

I was studying liberal arts only because that was what the college program required before I could major in geology then specialize in paleontology - I was still fascinated by dinosaurs. My parents' financial situation was strained. They were building a new house that was bigger and grander than the one I grew up in, and my father bought a brand new Coupe de Ville whose price tag I never saw. Add the three-week-long Caribbean cruises, and it is easy to understand why their money was tight and why they had none leftover for my education.

I was never an exceptional student so I couldn't meet the requirements for a scholarship. When my parents balked at continuing my support, my only choice was to get a job in order to put myself through school. I couldn't work part-time at night and continue my classes during the day. For me, learning was a full-time struggle that consumed my every moment.

I didn't have far to look for a job because my father was willing to hire me as a helper at minimum wage. My plan was to skip a semester then pick up classes where I had left off. That plan was dashed when I got caught in the draft. I was horribly depressed over the prospect of being taken away from my friends and familiar surroundings and thrust into the vulgar, brutal, insensitive world of the United States Army.

Basic Training

Circumstances forced me to grow up fast, although I seldom kept up with the pace. I was still a child at heart, but maturity is not a prerequisite to enlistment, as anyone knows who has ever dealt with Army officers and NCO's.

Basic training was a joke, jungle warfare school a farce. In basic training we spent an inordinate amount of time polishing boots and belt buckles (then low-crawling through the mud), standing at attention in the hot Georgia sun, learning march steps (which are never used in war), singing bawdy songs in cadence, practicing for parades (I was never in one), raking dirt into groomed parallel rows, presenting arms, gulping food, making beds, dusting and waxing and doing other general housekeeping chores, and organizing footlockers so that toiletries were arranged on the shelf with absolute geometric precision.

Even more adolescent was the system of military etiquette. Imagine grown men saluting each other and pretending to be Boy Scouts with a chest full of merit badges, or acting like arrogant whip-wielding masters to some people, then posturing like slaves to others. A bigger bunch of analretentives I have never met in my life. I found military life demeaning, disparaging, and distasteful. Perhaps "dehumanizing" is a more descriptive word.

I surprised myself more than the cadre when I first fired a rifle down the shooting range. I was a natural. My groupings were tight, and distant targets fell like leaves in autumn whenever I got them in my sights. I missed the company marksmanship trophy by a single round.

My delayed physical development ended coincidentally in the months prior to induction, so that after eight weeks of intense physical training I sprouted from 155 pounds of feebleness to 170 pounds of strength and stamina. I won the physical fitness trophy with a nearly perfect score.

These achievements might lead one to believe that I was a model soldier in the making. Nothing could be further from the truth. I turned down officer's candidate school because I perceived it to be a path toward irretrievable uniformity and the stifling - perhaps the destruction - of character and individuality. I was desperately afraid of losing my sense of identity in the military maw. I viewed the army as a hideous manifestation of George Orwell's *1984*.

I spent more time on KP than most trainees because of quirks in my personality which contrasted sharply with military protocol. I was often assigned to duty and details while others were given time off. These punishments were administered because I questioned authority whenever it was exceeded or misapplied, argued constantly over senseless or impractical demands, and found fault with the system.

For example, our commanding officer wanted to suck up to the post commander and demonstrate unilateral trainee patriotism by having every man in his outfit purchase savings bonds from money to be deducted each month from the trainee's pay for the duration of his enlistment. I declined the offer to buy not because $66 a month did not go very far as it was, but because I thought I could invest my money more wisely in other ways. The amount leftover after deductions would not satisfy predicted short-term needs, and the rate of return was too low and too late in coming. It was obvious to me that the CO knew nothing about investment leverage.

After a week of daily announcements about the CO's grandiose scheme, the platoon sergeants became hostile and abusive over the lack of complete support. Threats of retaliation predominated and abstainers were given extra watches at night. (I used the time to take care of personal chores.) Finally, the last dozen hold-outs were summoned to the CO's office where he gave his final pitch: if he did not obtain one hundred percent participation he would cancel all weekend passes for the duration. Half gave in, but the attempted intimidation resolved me more than ever to resist his manipulations. The CO then marched us to the parade ground where the rest of the men had been standing at attention awaiting results. Those of us who still refused to comply with his demands were lined up in front of the company. The CO announced to everyone that absolutely no passes would be issued until those of us holding out were convinced to change our minds.

This blatantly obvious tactical maneuver conveniently shifted the burden of convincement while adding immeasurably to the pressure that could be brought to bear. The threats increased, but now they came from my bunkmates. I explained patiently how they were being duped, but they could not be made to understand that the responsibility for denying passes lay solely in the hands of the commanding officer and could not be delegated to me.

On Friday afternoon, two hours before the usual time for the issuance of weekend passes, we half a dozen recalcitrants were hauled before an irate commanding officer. He screamed, he yelled, he cajoled, he threatened in a display of stereotypic childish tantrum. We remained defiant. Eventually he yielded. Everyone in the company got a weekend pass - except for the hold-outs. I spent the weekend scraping and painting buildings during the day and doing orderly duty at night. What any of this had to do with preparing a boy for war, I don't know.

Because of this and similar incidents in which I repelled indoctrination, I was considered to have a "poor attitude toward military service." Under the circumstances of determination, I didn't know if that was good or bad. I suppose it depends upon perspective. Today I'm inclined to think that it was better than the alternative.

Jungle Warfare School

I received jungle warfare training at Fort Polk, Louisiana, a real-life hell hole known as Tiger Land. It was commonly acknowledged that the bayou country surrounding the base was more barbaric than the jungles of Vietnam. I agree. So did President Truman when he expressed appreciation to the troops after World War Two (paraphrased from memory): "I want to thank all those who fought valiant-

132 – The *Lusitania* Controversies

ly in the war, and those who were stationed at Fort Polk, Louisiana."

The bayou was a perfect setting for jungle warfare training: part hilly forest, part lush swamp, largely tangled undergrowth, rugged, and inhabited by plants and animals leftover from man's primitive beginnings. Ants grew to the size of mice. Hairy spiders grew larger than the ants and waged furious battles against them. Wild boars with nasty-looking tusks roamed savagely across the land. Venomous snakes of every description were so prevalent that we were issued snake bit kits and given snake identification reviews every other day. Add armadillos and sergeants with a caveman mentality, and I was reminded so much of *The Lost World* that dinosaurs would not have appeared out of place. The only good thing going was that no one was shooting at us.

Here was little ceremony or wasted pomp and circumstance. We drilled in nothing but the weapons of war: the M-16 automatic rifle, machine guns, grenade launchers, the light anti-tank weapon called LAW (equivalent to the bazooka of World War Two), the Colt .45 automatic pistol, and, for some, mortars. I did not receive mortar training, but I qualified as expert in all the other weapons.

We practiced hand-to-hand combat and bayonet fighting, both fixed and unfixed. Small unit tactics were emphasized. We spent nights alone in the jungle on escape and evasion courses, and endured a couple of week-long field maneuvers of company strength with daily platoon-sized excursions into regions of unimaginable remoteness. Mock bamboo villages including fake booby-traps had been constructed deep in the jungle to complete the simulation of conditions to be encountered in Vietnam.

Even here, however, military intelligence reared its ugly head. Part of the simulation process was to carry only the gear that we would bear in actual combat. This meant no tents, sleeping bags, or jackets and warm clothing. When the nighttime temperature dropped to near freezing, we were wearing jungle fatigues with no other protection from the elements except a blanket.

Because we were not issued live ammuntion for maneuvers, attacks from wild boars were fought off by firing blanks in their direction. More than a few snakes were killed in this manner by the muzzle flash. One man nearly died from the sting of a scorpion.

I helped carry the injured man up the hill we were "guarding." Within five minutes of the poisonous sting his foot swelled to twice its normal size. A jeep took him to the base hospital where for four days he lay groaning, packed in ice to reduce a temperature of 104°.

During guard duty the next day I had the misfortune to sit on a scorpion nest. I didn't know it till I felt something squirm under my leg. With cold chills coursing along my spine, I stopped breathing and did my best to imitate a chunk of granite. The scorpion worked through the loose folds of material and emerged between my legs with its miniature claws quivering. It was a delicate-looking creature only half the size of my pinky and nearly white in color. I didn't dare move. It crossed the short distance between my legs, crawled under my other leg, then settled down for a moment in the warmth provided by my flesh. I couldn't even call for help lest the merest motion trigger its deadly defense system. About a week and a half later it crept out from under my olive drabs and slowly scrabbled off across the moss-covered rock. Not until it was five feet away did I draw my first breath. Then I jumped to my feet and searched frantically for its brethren. None was around. I went after the scorpion with my canteen cup, managed after a few tries with a long stick to steer it over the lip, then carried it about a hundred yards into the jungle at the base of the hill and flung it into the bushes, figuring that it would never find its way back to our guard perimeter.

Classes missed due to KP or illness could not be made up later. Low grades didn't count, and for most subjects tests were not even given. Those who strived to do their best did so on their own initiative. No incentives were offered nor rewards promised for excellence. There was no obligation to learn.

I was thrown in with a wild bunch of teens who were street smart but woodcraft illiterate, and proud that they had broken all ten commandments and the Boy Scout mottos. They were a tough lot of characters used to inner city conflict - the kind of guys you wouldn't want to meet in a darkened alley. They would gladly brawl over the slimmest of slights however imperceptible. Yet most of them were likeable and friendly toward those they did not consider adversaries.

The largest proportion thought of themselves as "heads," which they explained to the only unsophisticate in the outfit - me - meant "potheads" or "grass smokers." Since I didn't understand the meaning of grass in any context other than what grew in lawns to be mowed, they had to explain that too. Marijuana was available in unending supply. When they smoked all the grass they brought with them, more arrived via the postal service in plain white envelopes with no return address.

On the nights that we got to sleep in the barracks, the "heads" slipped out into the nearby forest and smoked themselves into a stupor. If they didn't come back on their own - the usual case - I went looking for them and tried to convince them to return before bed check, so they wouldn't get into trouble. They took a lot of convincing. Often I found them lying stoned on the ground incapable of motion. I dragged them to their feet and led them back to their bunks in a train.

A great deal has been said about soldiers picking up the drug habit during their tour in Vietnam. The way I saw it was just the opposite: soldiers already had the habit from their free-wheeling civilian lifestyle, and carried it with them to where marijuana was just as readily available.

Violent urban upbringing did not prepare these guys for the rigors of outdoor life. For example, one night we were dropped off by squads at an unknown location in an endless swamp. Each group of six or eight had a different starting point and was given a different compass heading to follow. The goal was to bushwack through five or six miles of thick trackless territory and to arrive at a specific point, where a truck would be waiting to drive us back to the barracks. We had to count steps and maintain course while beating through the underbrush, winding through dense forest up steep hills and down, fording shallow creeks, and wading waist deep through muddy marshes - all in total darkness. I loved it.

I was arbitrarily elected leader by the members of my squad because none of the others knew how to estimate the distance traveled or to read a map or lensatic compass. These big city ruffians wilted in the wilderness. One man screamed and fell back over a branch when a rabbit darted out in front of him. Another dropped his rifle while jumping across a creek. We had to slide down the embankment and drag for it with long sticks which I cut with my bayonet in order to fish it out. Two got temporarily lost in the swamp when they lost physical contact with the man in front. The darkness was absolute, and because the lush vegetation was impenetrable in spots they could not follow the direct route led by sound. I splashed my way back to the point where I estimated they had gotten separated from the pack, found them from there, then led them on by the hand.

Despite these diversions we were the first squad to reach our destination. The sergeant in charge could hardly believe that we had done in three hours what was expected to take six. There was no way to cheat. He congratulated us, then ordered a driver to take us back to camp - some forty miles away over rough dirt roads - to have a well deserved night's sleep.

We were allowed to stay in bed as long as we wanted after this particular exercise. I moseyed out for breakfast about the middle of the morning. The menu offered whatever I chose to have, and it was cooked to special order - even steak and eggs. Because the barracks had been deserted I expected the mess hall to be full, but it was not. Only half the men had made it back so far. Most of the remainder drifted in throughout the day with haggard looks on grimy faces and exhaustion in their eyes. Many had cuts and bruises that required medical attention. They all had horrid tales to tell.

By evening nearly everyone was back but still a few were missing. Jeeps were dispatched for night patrol, equipped with loud speakers and microphones as well as food and water. (We had been issued no rations.) A few more men were scooped up after dark, suffering horribly from dehydration. The next day the last of the stragglers trickled in after escaping from the swamp and reaching dirt roads, which they followed to an asphalt. Some were picked up by the state police and brought back to the base. Others hitched rides with local Cajuns who took them to the nearest town where they were able to catch a bus. The last survivor did not return till forty-eight hours after drop off. He rolled up to headquarters in a taxicab, then had to borrow money to pay the fare.

These were the men being sent to fight the war in Vietnam.

INTO THE JUNGLE

No one flunked jungle warfare school. Nine weeks of accelerated training was all the preparation received to engage a military force whose soldiers had been born in the jungle, raised in the jungle, and taught since childhood to fight in the jungle. In this unequal contest of amateurs opposing professionals, the amateurs had the advantages of unlimited resources and absolute control of the air against the stealth and concealment of the professionals.

There was no front line in the accepted meaning of the term. The entire Republic of South Vietnam was a vast battleground in which two great armies intermingled like grains of salt and pepper on a jungle salad. Saturation bombing, napalm strikes, and defoliation offensives were large scale tactics as inappropriate under the terms of engagement stipulated by enemy strategy as burning down a house to exterminate a nest of wasps clinging under the eaves, or using a chainsaw to excise a pea-sized cancerous tumor.

The problem we faced in the bush was that for the most part the VC and the NVA declined to fight full-scale battles. Nipping at our heels like Chihuahuas that were too fast to kick, Vietcong guerrillas and regulars of the North Vietnam Army played a deadly game of snipe and run that was difficult to combat. This is why small unit actions of company strength were more effective than large sweeping assaults.

I was part of a canary campaign. In coal mines with long tunnels where deadly gas accumulates, a canary warns the miners of poisonous vapors by keeling over in its cage. In the jungle, we scouted hamlets and villages and tried to locate the enemy. If we made contact, reserves were choppered in to join the firefight. We were the sacrificial canaries.

CLOSE CALL WITH DEATH

I felt a clap next to my ear and almost simultaneously heard the retort of a rifle on my left. The bullet passed so close to my head that the collapse of air into the vacuum of the projectile's wake created a disturbance that brushed my lobe. In a single fluid motion I spun, dropped to one knee, and fired at the back of the sniper as he vanished into the jungle. During that fleeting glimpse of olive drab I had snapped off five fast rounds.

As experienced troops, my fellow men-at-arms dropped and aimed their rifles at the jungle close to the path, each alternate man facing the opposite side. No other targets presented themselves, no enemy fire ensued.

My platoon leader raced back along the path in a crouch. In a fierce hush he demanded, "Who fired those shots?"

In a few curt sentences I explained what happened and pointed to where I saw the sniper retreating during my brief but perfectly clear shot. In my ignorance I asked, "Shall I go after him?"

For a long moment the peach-fuzz lieutenant stared thoughtfully into the jungle. "No. That's just what they want us to do." The side of the path could be lined with hidden punji stakes smeared with feces, or the jungle floor could be sown with pressure actuated booby-traps, or the trees could be strung with trip wires connected to explosive charges, or an ambush could be waiting for us just inside the tree line. "Move on. Keep your eyes pealed. Be on the alert."

Not until the end of the day did I have time to reflect on the occurrence, and to realize that my life had been only an inch away from death.

BULLETS AND WATER

The temperature was 110° in the shade and we were in the sun. Sporadic fire erupted from the bamboo thicket just inches in front of my face. I ducked low behind the foot-high dike and conserved my ammunition. I had already tossed all my grenades except for one I always kept in reserve. We'd been pinned down for a couple of hours without being able to advance. I rolled on my side to take a swig from my canteen. It had been empty for an hour but half a dozen times I had tried hopefully to find another drop with my tongue. It was still empty. I crawled a dozen feet to the left to where my squad leader lay pressed flat against the ground. His canteens were empty, too. We were parched.

I remembered seeing a flooded paddy half a mile back along our approach. I suggested that I collect the squad's canteens and go for water. He thought it was a good idea. I crawled along the dike and gathered canteens, all except for the two-quart plastic bladder that the machine-gunner carried on his lower back: it had been shot full of holes and both the back and front of his pants were soaked as a result. I strung a spare shoelace through the cap retainers, signaled that I was ready, and low-crawled fast while my team members maintained covering fire.

A few enemy rounds got past the barrage, but none came close enough to matter. It was like crawling over the confidence course in basic training except that the live rounds fired overhead were intended to kill, and I didn't have to spit shine my boots that night. When I got far enough away I climbed to my feet and scrambled in a crouch for a quarter of a mile, then walked along easily but fully alert. I was alone, and the enemy was definitely nearby.

The paddy was rank and so coated with scum that it looked like a golf green with a par of three. I stepped gently into the water so as not to stir up the bottom, then parted the green foul film with my hands. I ignored the drooling slime I saw hovering above the mud. I completely submerged the canteen so the siphon effect would not draw dirty water off the surface. When the canteen was nearly full I plunked an iodine tablet through the mouth, shook the canteen with my thumb over the top, then drank the whole quart at a single draft before the tablet had time to dissolve.

I don't think many people have a true appreciation for the power of thirst. I emptied the second canteen in two quick drafts. Never in civilian life have I been able to duplicate the feat. After filling the canteen for a third time I had to move on to a new location because fat ugly leeches, sensing the heat from my hands in the water, were converging from various directions and arched only inches away from my flesh.

When I returned to the squad with the water I was the hero of the day. The firefight was still in progress. So was the war.

No Rest for the Wary

One day our squad paused for a rest in the shade while the rest of the company advanced to the perimeter of a village we happened to be passing on the way to our primary objective. Standard operating procedure required that we conduct a cursory examination for enemy sympathisers.

The general definition of "sympathiser" was anyone who supplied the enemy with food, shelter, supplies, or information of military value. The specific definition varied according to viewpoint. To us, sympathisers were unarmed peasants who lent aid to the VC or NVA under threat of death or rape, in which case they were slain and the village destroyed by American troops under orders. On the other hand, unarmed peasants who offered no resistance to American incursion or who otherwise minded their own business while American troops passed by, under threat of death or the destruction of their village, were executed by the VC or NVA for sympathising with the invasionary forces. By these contradictory formulations, every Vietnamese peasant was characterized as a sympathiser by one of the warring factions - a lose-lose situation no matter how they looked at it.

I was having fun knocking coconuts out of a tree - one use for an entrenching tool that was not covered in the training manuals. I leaned my rifle leaning against the trunk in order to have both hands free so I could chop a hole in the husk with my bayonet and drink the cool refreshing liquid within. Suddenly there came the crackle of gunfire nearby. I grabbed my rifle and charged into a path where forty feet away a VC guerrilla was climbing out of a spider hole, spraying bullets. He was gunned down from three sides before I had time to react. After that, I never let my rifle out of my hands again.

A few minutes later I found a storage bin the size of an outhouse that was half full of rice. I poked through it with my bayonet and to my horror uncovered a booby-trap from which loose wires dangled. For several seconds I stood stock still - waiting for the blast to take off my hand or my face - until I determined that the device had not yet been armed. The village was definitely "hot," yet no reprisals were taken.

Search and Destroy Mission

Another day we entered a village that was occupied only by women, children, and leathery old men. We found no signs of the enemy, no concealed weapons, no buried booby-traps - nothing that would indicate collusion with the VC or NVA other than the absense of virile young men. Hueys arrived with 50-pound wooden cases of explosives which we distributed throughout the ville: one inside each bamboo hooch, two inside each mud-walled bunker. We ran out of dwellings before we ran out of explosives. Rather than save the six extra cases for another job, we placed all three hundred pounds of explosives on the dining room table in a French-built stucco house.

The peasants were evacuated on the hueys that brought the explosives.

I asked for permission to release the livestock before the village was destroyed. In a rare moment of sanity and sensibility, permission was granted. I ran from stable to stall to henhouse letting out pigs and goats and chickens till the hardened dirt roadways were choked with squeals and bleats and clucks as well as the curses of American soldiers who disdained the animals running amok.

We strung roll after roll of primacord throughout the compound until every case of explosives was wired together on one continuous string. The troops gathered outside the village perimeter while the final splice was made. When the electrical contacts were touched, the detonator ignited the primacord which burned at 27,000 feet per second - that's five *miles* per second - and set off all the explosives with near simultaneity. The entire village blew up in one titanic roar of flying dirt and debris. There was nothing left of the stucco house but the splintered concrete pad.

Cooking off Grenades

One ville resisted our reconnaissance with automatic weapons fire. The shooting was by no means hot and heavy, but it was consistent. We beat through the brush past the first row of now-vacant hooches, behind which fortified bunkers protected the central square of the ville and impeded our progress by the possibility of their occupation. Each squad singled out a bunker to attack. With a burst of covering fire I ran across the open space and past the doorway of the hooch, and plunked down alongside the thick mud wall that was every bit as strong as concrete.

I yanked a fragmentation grenade off my web gear, pulled out the pin with my teeth (my rifle was in my left hand), and, clamping the release lever tight with my thumb, waited for the signal from Yawn. Once the lever is released there's a four-and-a-half to five second delay before the grenade explodes - enough time for a nimble adversary to catch it and toss it back.

In close quarters such as this I had to "cook off" the grenade. Yawn signaled. I let go of the spring release, stared at the live grenade for precisely two seconds, swung my arm back and then forward, and let go of the grenade on the count of three. The grenade ricochetted twice then went off with a loud detonation that shook the air and filled the bunker with a deadly shower of shrapnel.

Without waiting for the dust to clear I ducked into the bunker doorway and crept inside, leading with my rifle. The bunker appeared to be unoccupied. Quickly I flipped over a sheet of plywood covering part of the floor - it might be a trap door for a spider hole or tunnel complex. I saw solid earth crawling with large red ants that immediately ran up my arms and legs, biting angrily. Better poisoning from formic acid than from lead.

Overkill

Not all the bunkers were vacant, and soon the bodies were being dragged out, sometimes a piece at a time. Crouching by a spider hole which my platoon sergeant had eagerly jumped into, I leaned back to avoid the body parts that he tossed out so irreverently: a foot, a hand, a bloody glob of something that was unrecognizable.

One VC guerrilla lay on the ground on his back. From the looks of him a grenade had gone off in his lap: shredded organs lay exposed in a gory hole that stretched from crotch to rib cage. His manhood and both hands and feet were missing. A medic knelt by his side and - I could hardly believe it - *felt for a pulse!*

I thought the medic's action was insane until he announced softly, "This man's still alive." He let go of the truncated forearm.

My platoon sergeant put the barrel of his M-16 to the guerrilla's temple and pulled the trigger, splattering bone and brain everywhere. "He ain't alive now."

THE IRONY OF BATTLE

On another occasion this same sergeant, whose name I've forgotten, demonstrated a different kind of machismo. We were pinned down for hours in a stand-off firefight in which neither side could advance or was willing to yield. We crawled on our bellies through nearly impenetrable jungle in order to close the action. I never saw the enemy, I just fired into the bush where I thought I heard movement or the cracking of a rifle. We were close enough to the enemy to exchange grenades, but we had to be careful that the grenades we tossed didn't bounce back off the boles of the densely packed trees.

After a lull in the action came the order to fix bayonets for hand-to-hand combat. I heard a warning shout in front of me. When I looked up, I saw my platoon sergeant with an enemy grenade bouncing between his chest and his hands. I leaped out of his way - everyone did - as he ran by and slammed the grenade down into a bomb crater. It was a dud.

The shooting picked up immediately. I found myself on the ground behind the bush I had jumped over. I low-crawled around it and reached a position of relative safety behind a slight rise of dirt, while overhead, bullets chopped the leaves into confetti. I emptied one magazine and loaded another. Then came the order to advance. I stood in a crouch and instantly felt a sharp pain in my ankle. I fell painfully. I thought I must have been grazed by a bullet or chunk of shrapnel. I continued to move forward on my knees until the enemy melted through the jungle and vanished.

I removed my boot and examined my foot. It was already swollen enormously, but there was no blood. I had twisted my ankle when I had hit the ground while avoiding the sergeant with the dud grenade, but in the heat of battle hadn't felt the snapping tendons. The medic had no elastic bandages, so I pulled on another sock and laced my boot extra tight. Then I limped five miles through jungle and rice paddies to our company's perimeter of sandbag bunkers. There I got an elastic bandage for my ankle. We were so short-handed that I didn't even get a day off. I continued to go on daily patrols with my swollen ankle bandaged.

Vignettes in Perspective

In another action, a guerrilla in his spider hole was surrounded by my squad and was asked to surrender. His reply was a grenade - which he clutched to his stomach till it went off. He died instantly. What propoganda did the North Vietnamese furnish their troops about the treatment received by prisoners of war at the hands of American captors. How much of it was true?

One firefight got so fast and furious that I grabbed the machine gunner's M-16 and fired both his and mine from the hip on full automatic, while he poured lead into the jungle until the machine-gun barrel turned cherry red. Eventually the enemy ceased firing. We broke off contact without a chase, for fear of ambush or booby-traps.

In the Central Highlands we kept on the go and dug new fox holes every night, with claymore mines set beyond the perimeter after dark as a first line of defense. In the coastal lowlands we stored supplies in an abandoned rice paddy which we used more often than we should even though it commanded a wide field of view and lay conveniently close to our operational area. A few men guarded it during the day while the rest of us went out on patrol. We never had trouble finding it after a long trudge in the bush, and neither did Charlie. I often wondered why the enemy didn't attack it during the day when it was so lightly defended.

Charlie never did, but not a night passed without harassing sniper fire. Each barking streamer of tracers sent a hundred men scrambling for sandbag bunkers, reminding me of a Decatur Road getaway after the police were sighted - or was that drag race scene part of my vivid imagination?

Crack Shot

According to intelligence, the NVA was massing in our area for a major assault, and there existed the possibility of being overrun. That night, while standing observation post, I witnessed the predicted action begin but against our neighboring company. Across several miles of open paddy came the cacophany of gunfire, mortars, and artillery as the attack proceeded, and the flashes from incoming rounds. I detected the presence of a dive bomber by the light of its tracers as it spewed a solid stream of lead beyond the perimeter of the outfit it was protecting.

I felt some relief in presuming that Charlie's forces were spread too thin to attack two positions at once, and that while our fellow soldiers were dying nearby, we should survive the night unscathed. It might be our turn next.

While standing observation post and watching the distant show, I thought about the deserted hooch I had spotted before dusk. Our defensive circle was about two hundred feet across. Along one quadrant the jungle began only one hundred feet from the perimeter. The other three quadrants faced unworked paddies miles in extent. Patches of jungle broke up the monotony and blocked the view of a curved horizon. The deserted hooch sat secluded on the edge of a small patch of jungle, and one side had a clear field of fire facing our temporary encampment.

Our nighttime routine was to launch a grenade about every half hour, in order to let the enemy know that we were on the alert. The whumping of grenades did further service by keeping awake the men on guard. Whenever someone decided that enough time had passed to launch the next grenade, the others on guard did the same. It might seem that this constant but sporadic explosive uproar - from grenades, gunfire, mortars, and claymores - and the agitation consequently aroused, would make nighttime slumber impossible. On the contrary, I found it difficult to remain vigilant even with a starlite scope glued to my eye. I was constantly drowsing despite the potential proximity of death.

With my nerves on edge from the attack in progress against our fellow troops, with the possibility of imminent attack ever on my mind, the more I thought about that secluded hooch the more uneasy I became. After a lengthy period of silence I shifted the grenade launcher in my lap. It was an over-and-under model, secured beneath the barrel of an M-16 but sharing a common stock. The rifle and the grenade launcher had separate triggers, not unlike a two-barreled shotgun.

In the pitch blackness of night I couldn't see the hooch. I extrapolated its range and bearing from memory. I estimated the distance at four hundred fifty yards - more than a quarter mile - elevated the barrel accordingly, and adjusted my sights for windage. The sleek-skinned grenade left the tube with a whoosh, lobbed like a badminton shuttlecock, and detonated with a dull thud - followed immediately by flashes of light and long tongues of flame, then by a massive explosion which sent blazing thatch and bamboo splinters high into the nighttime air.

Incinerated debris showered down for many minutes, like a movie in slow motion. The remains of the hooch burned for hours with occasional bursts of brilliance. A person used to viewing Hollywood pyrotechnics might think such a discharge normal, but in reality fragmentation grenades are anti-personnel devices that blast shrapnel through flesh. In this case, a secondary explosion triggered by the sparking of metal on metal must have accounted for the titanic blast that followed.

Had I hit an enemy ammunition dump? We retreated the next morning without investigating the hooch.

Sleep Deprivation

One night I slept through an artillery attack.

I admit to being aroused by the first incoming round, but because it exploded a hundred yards from where I lay on the ground, I merely opened my eyes without otherwise stirring. We had no tents or blankets, just mosquito mesh to sleep under. With ostensible detachment I watched men dash madly for shelter. I was determined not to quit my bed of earth until the rounds were walked close enough to be a sure danger to health. I luxuriated in complacency born of the youthful sense of invulnerability and the practical necessity that was demanded by fatigue.

The jungle erupted with a brilliant flash of orange. Leaves were shredded from trees and torn limbs fell. When the next round landed a hundred feet closer I heard shrapnel whistling through the air, followed by thumps and sizzles as jagged, fist-sized chunks of the projectile casing struck the damp sandy soil less than twenty feet from my head. I calmly detached my helmet liner from my steel pot - which I had been using as a pillow - put my head down on the helmet liner, then balanced the steel pot over my face. I pulled my rifle tighter against my chest - I always slept with my hand on the trigger guard so I wouldn't have to find it in the dark - and dozed off.

I opened one eye at the concussion of the third round, then knew no more till dawn when my squad leader kicked my leg with his boot. I had slept through another thirty minutes of artillery fire, none of which came close enough to arouse me. My team members saw me lying on the ground but thought I was dead. I shocked the heck out of them when I yawned and nonchalantly asked about breakfast.

One who has never been subjected to the exhaustion of constant combat might find it difficult to accredit such a story. In the tranquility of an ordered and sane society, people are often roused by the slightest sound - the dripping of a faucet or the creak of a board. But in the bush we were so spent from the long daily patrols, the frequent intense action, the emotional strain of fighting, and the lack of unbroken sleep, that a state of chronic lassitude pervaded the mind.

Failed Leadership

Sometimes I carried tear gas grenades and a gas mask in addition to my usual gear and ammunition. As I forced my way through dense jungle foliage a branch pulled open the snaps of the canvas case, which I wore around my neck on a strap, and the gas mask fell out unnoticed. Not until we stopped for a break did I realize that the mask was missing. I told my squad leader about it, and Yawn sent me to report it to our platoon leader who, with the other lieutenants, was talking with the company commander. The captain ordered me to go back and find it, not because it was so valuable, he said, but because he did not want it to fall into enemy hands and be used against us. My platoon leader detailed two men to accompany me.

After half a mile one of the men refused to go any farther. He was scared, and I don't blame him. Without the comfort of numbers our situation, should we make contact, was perilous. No one knew where the enemy was. He could be following us, he could be surrounding us, he could be ambushing us. After another quarter mile the jungle opened onto a grassy plane which the second man refused to cross. It was too open, he said, and we would be too easy a target for a sniper. I argued, but the most I could get from the man was a promise to cover me and to wait ten minutes for my return.

I crouched down low and ran across the field to the tree line opposite where the company had only recently pushed its way through. Within five minutes I found the gas mask lying in the underbrush on the side of our bushwacked trail. I tucked it into the case and fastened the snaps. Back at the edge of the field I waved, and was relieved to find that my reluctant companion was still waiting for me. He signaled for me to cross. The other man had returned on his own.

I reported the success of my mission to the captain. He said, "Let that be a lesson to you." He called for the troops to march on.

Was he truly concerned about Charlie obtaining a gas mask, or did he want to teach me a lesson to be more careful with army equipment? In retrospect, I think he made a poor tactical decision to risk the lives of three men simply to prove a point.

Body Count

The fighting in Vietnam was different from that of conventional modern warfare in that there were no strict lines of confrontation. Nor was there any way to distinguish territory dominated by friendly forces from territory dominated by the Vietcong. The country belonged to whomever occupied the land at a particular moment in time. Dominion changed hands according to a semi-circadian cycle: American troops and the Republic of Vietnam held tentative command by day, while Charlie controlled the night. All was enemy territory.

Against this backdrop of exchanging supremacy there was no way to gauge the progress of the war. So the United States established a scoring system of attrition called the "body count." According to this scheme, every dead enemy soldier brought the friendly forces one step closer to victory. But the system got out of control when the definition of "soldier" was expanded from arms-bearing warrior to include unarmed civilians called "sympathizers" and, all too often, woman and children caught in the line of fire. This ultimately led to a pathology for which, to American shame, the Vietnam conflict will forever be remembered.

Photographic Insert – 145

Lusitania in Clydebank Dock.

Above: A postcard view of the *Lusitania* in the Clydebank dock. Notice how she towers above the other vessel. (Eric Sauder Collection.)
Below: A crowd gathered on the quay. (From the author's collection.)

146 – The *Lusitania* Controversies

Two postcard views of the *Lusitania* in Liverpool.
(Both Eric Sauder Collection.)

Photographic Insert - 147

Above: The *Lusitania* leaving on her maiden voyage.
Below: The mammoth hull towers above curious onlookers.
(Both courtesy of the National Archives.)

148 – The *Lusitania* Controversies

Above: The *Lusitania* in all her glory.
Below: The distance at which the *Lusitania* normally passed the Old Head of Kinsale.
(Both Eric Sauder Collection.)

"LUSITANIA" PASSING OLD HEAD OF KINSALE

Photographic Insert – 149

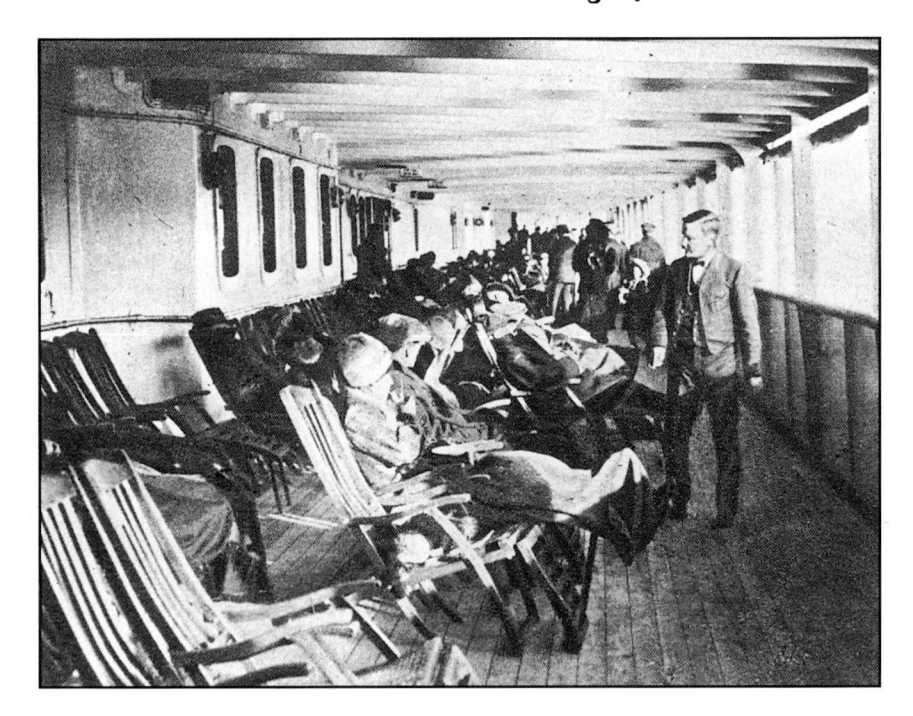

Above: Passengers enjoying the
wide promenade deck.
Below: The spacious lounge.
(Both Eric Sauder Collection.)

150 - The *Lusitania* Controversies

Two views of the first class dining saloon.
Above: Post card view. (Eric Sauder Collection.)
Below: Upper and lower saloons and the vaulted
ceiling. (From the author's collection.)

Photographic Insert – 151

Above: Private dining room in the regal suite.
Below: The utilitarian third class dining saloon.
(Both from *Scientific American*.)

152 – The *Lusitania* Controversies

Above: First class stateroom.
Below: Second class stateroom.
(Both Eric Sauder Collection.)

Photographic Insert – 153

Right: Bedroom in the regal suite.
(From *Scientific American*.)
Below: Third class cabin, representing less opulent accommodations that were not advertised in brochures or other promotional literature.
(Eric Sauder Collection.)

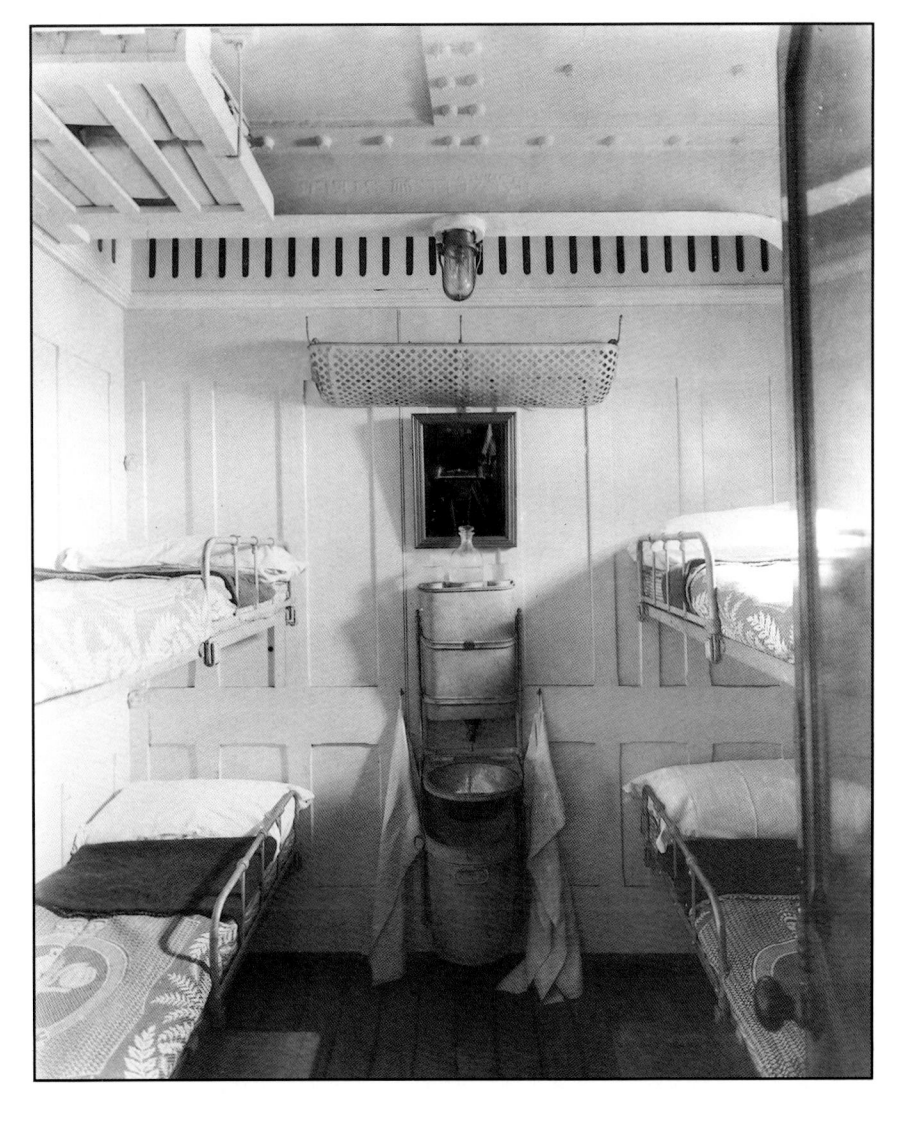

154 - The *Lusitania* Controversies

Above: The docking gang at work on the bow. (Courtesy of the National Archives.)
Below: The stokehold between boilers is where the black gang shoveled coal into the furnaces, creating coal dust in the process. (*Aquitania*, Eric Sauder Collection.)

Photographic Insert - 155

Above: The notice that was published in New York newspapers on the day the *Lusitania* departed for her final voyage appears adjacent to the liner's advertised schedule. (Eric Sauder Collection.)
Right: A bedraggled Captain Turner after his rescue. (Courtesy of the National Archives.)
Below: The *U-20*. (From the author's collection.)

156 - The *Lusitania* Controversies

This illustration, drawn by G. H. Davis, was published in the British newspaper *The Sphere*, May 15, 1915. (Eric Sauder Collection.)

Photographic Insert - 157

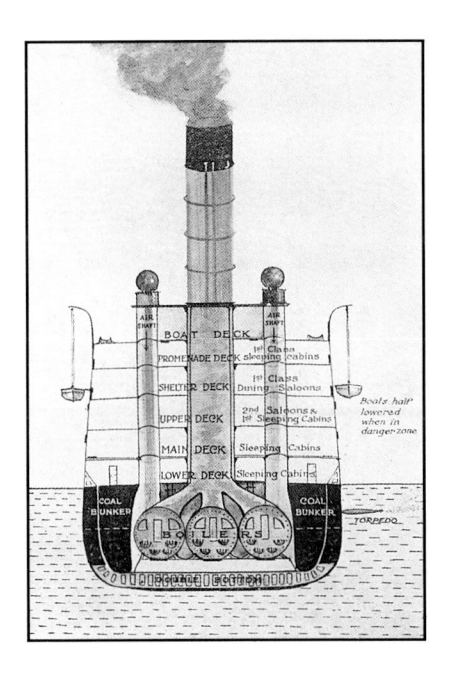

Above left: The bunkers with respect to the torpedo strike.
Above right: The dramatic cover of *The Sphere*.
Below: Another representation of the sinking.
(All Eric Sauder Collection.)

158 - The *Lusitania* Controversies

Queenstown teeming with the living and the dead.
Above: The anguished faces of survivors at the train station.
Below: Displaced survivors walking the streets. (Both Eric Sauder Collection.)
Opposite page: The deceased are transported to the cemetary.
(Both courtesy of the National Archives.)

Photographic Insert – 159

160 - The *Lusitania* Controversies

Above: The funeral procession passing through town.
Below: Local priests offiating the mass burial ceremony
at the Queenstown cemetary. (Both Eric Sauder Collection.)

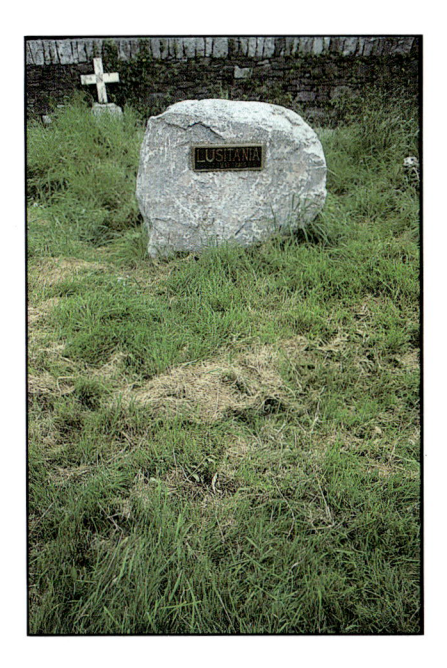

Photographic Insert – 161

Below: In 1994, weeds and grass grew around the large stones that marked the mass burial site. Beyond the stone stands the cemetary's perimeter wall, to which is secured a marble commemorative plaque.

Bottom: A close-up view of the plaque.

Above: One of the large stones marking the mass burial site, which in 1994 was overgrown with weeds and unkempt grass. Below: A nearby vault.

162 – The *Lusitania* Controversies

Queenstown's name has been changed to Cobh (pronounced "cove"). As shown in the photos above and below, in 1994 the *Lusitania* memorial was undergoing restoration and the town square was being repaved.

Photographic Insert – 163

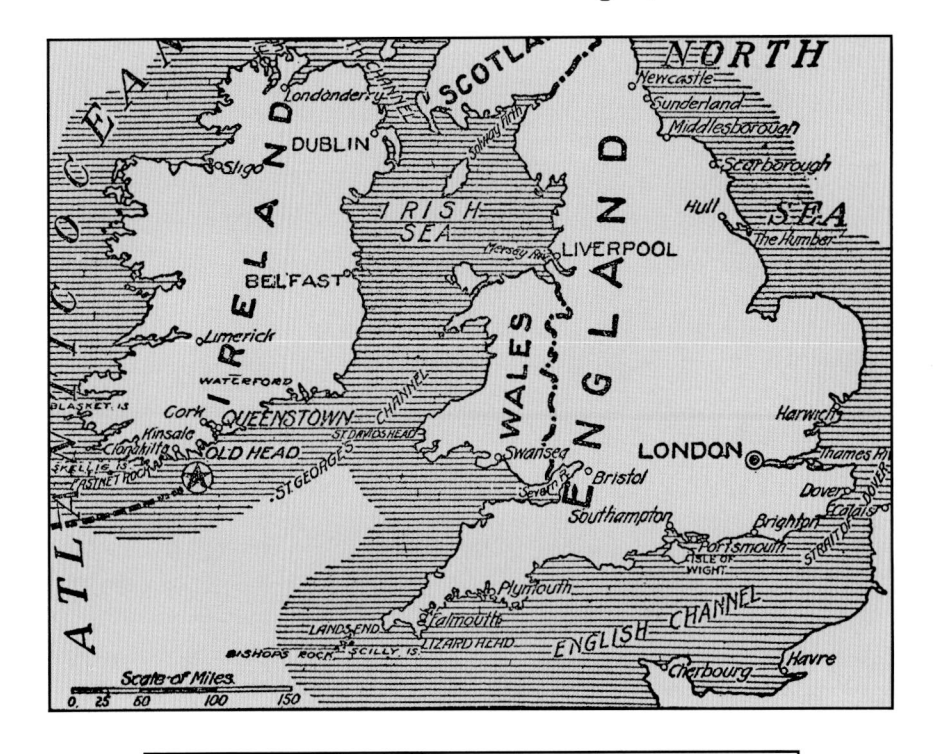

Above: The location of the sinking. (From the author's collection.)
Below: The lighthouse still stands guard on the Old Head of Kinsale.

164 – The *Lusitania* Controversies

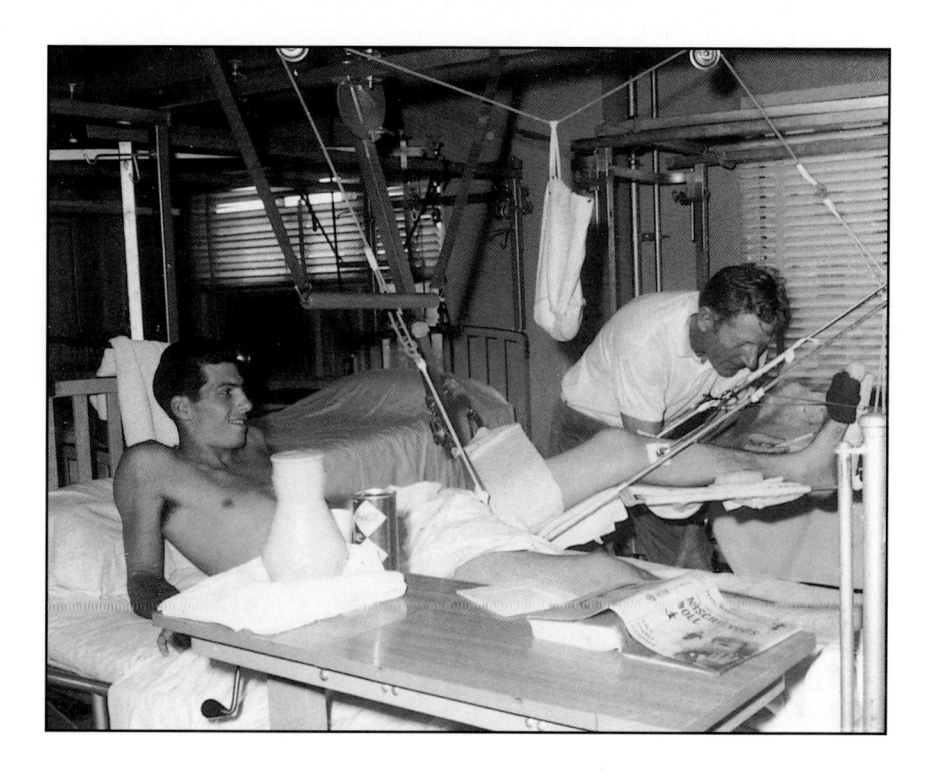

Photographic Insert - 165

Dramatis personae in the 1960's.

Opposite above: John Light, dour as always, stands in front of his diving crew. (Photo by Paddy O'Sullivan.)

Opposite below: Victim of war, the author spares a painful smile as Danny Kaye pauses to draw a happy face on his foot while visiting patients at 106th General Hospital, Kishine Barracks, Japan.
He added a ray of sunshine to an otherwise dismal existence.

Above and right: Mike de Camp, setting the stage for the future course of wreck-diving. (Both courtesy of Michael A. de Camp.)

166 - The *Lusitania* Controversies

Adventures in spelunking inspired my diving career.

Upper left: I exhaled all the air in my lungs in order to force my chest through a tight squeeze. (Snivley's Cave, Maryland.)

Middle left: Blue Hole, Virginia.

Lower left: Al Dubeck steadies the raft while Tom Gmitter inflates the starboard pontoon. The blackness beyond is Blue Hole's underground river.

Photographic Insert – 167

Wilderness water treks.

Upper and middle right: Carrying gear and canoe on the portage trail, Riviere Du Chef, Quebec.

Lower right: My partner and I have already run this series of rapids on the Mistassini River, Quebec. Two of the party watch from the bank as the other two negotiate the rocks and standing waves.

168 – The *Lusitania* Controversies

Other outdoor activities.
Upper left: I am climbing Long's Peak the hard way (in the Rocky Mountains, Colorado).
Upper right: The Crestone Needle, in the Sangre de Cristo Mountains of Colorado, which I climbed solo and without protection. For the 2,000-foot ascent I climbed straight up the face to the V-shaped fracture. The left side of the V proved too difficult unaided, so I retreated and climbed up the right side, forced a chimney below the summit, then angled left for the final assault over the lip of the vertical precipice.
Lower left: Jack Schieber negotiates a narrow ledge in the Grand Canyon, under the watchful eyes of his Smith family nephews and neice: Drew, Gwen, and Todd.
Lower right: Winter mountaineering on Mt. Mansfield, New Hampshire. Drew Smith is wearing crampons, Jim Murtha is wearing snowshoes. Intense cold, high winds, frosted eyelids, and occasional whiteouts offered challenges to meet and overcome.

Photographic Insert – 169

Above: The beginning of a friendship - the author and Bill Nagle holding the compass from the *Bass*.
Below left: Bob Archambault, the author, and Bill Hoodiman with the helm from the *Ioannis P. Goulandris*.
Below right: Jon Hulburt, one of wreck-diving's greatest innovators.

170 – The *Lusitania* **Controversies**

Above left: George Hoffman in the *Sea Lion's* wheelhouse.
Above right: Tom Roach celebrating his birthday on the boat.
Below: Big John and Little John. Left: John Dudas. Right: John Pletnik.

Photographic Insert – 171

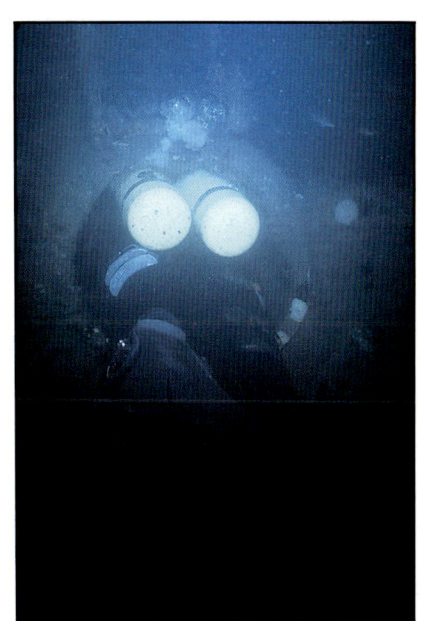

Top left: Tom Roach holding a couple of shipwreck souvenirs after a dive we made on the *Moonstone*.

Top right: Roach squeezing through a hatch into the *U-853*. On the way out, because of increased buoyancy, he ran into the bulkhead above the hatch. He had a few scary moments until I came back and showed him the error of his way.

Middle right: Mike de Camp with a name plaque he found in the wheelhouse of the *Varanger*, years after he first dived the wreck.

Bottom right: De Camp floating high and dry in his newfangled drysuit.

172 – The *Lusitania* Controversies

The dreaded bends.
Above: Danny Bressette takes a helicopter ride. Waving goodbye from the basket, he looks very much like the Wizard of Oz on his way back to Kansas by balloon.
Below: The helicopter hovers overhead as Bressette is winched up.
Opposite top: One of my rules of photography is "always get the shot." I took this shot from the supine position, despite partial paralysis and the awful fear of uncertainty. Looking down are Bill Hoodiman, John Starace, and George Hoffman.
Opposite bottom: I took this shot of the *Sea Lion* from the basket as it dangled below the helicopter. By this time my symptoms were considerably alleviated.

Photographic Insert - 173

174 – The *Lusitania* Controversies

Eastern Divers Association in its heyday.
Above: The *Bidevind* trip in 1973, aboard Ray Ettel's *White Star*. Wearig the tee shirt is Jan Nagrowski; behind him are Walt Krumbeck, Mike de Camp, Joel Entler, and Donn Dwyer; George Hoffman has his arm around John Starace; behind Starace are Danny Bressette and John Asqui; behind the author are Tom Roach and Jim Snyder; on the author's left are Bob Archambault, and behind him are Bill Hoodiman and Don Nitsch. Below: The *Andrea Doria* trip in 1974, aboard the catamaran *Atlantic Twin*. Kneeling from left: Joel Entler, Jan Nagrowski, Jim Snyder, and Danny Bressette. Standing from left: the author, John Asqui, Ray Bailey, Bob Archambault, Tom Roach, Donn Dwyer, Ron Burdewick, and John Starace. I am proudly wearing the "tourist" tee shirt that I was given for getting bent. Starace and I were dubbed "the Ford and the Ferrari."

Photographic Insert – 175

Above left: Norman Lichtman. Above right: Bart Malone. (Both by Gene Peterson.) Below left: A typically crowded east coast dive boat, with tanks, equipment, and people packed gunwale to gunwale. The most oft spoken phrase: "Excuse me." Below right: After rough seas, dive gear has reached the point of lowest potential energy. Notice the plastic trash containers: popular gear cases at the time.

176 – The *Lusitania* Controversies

Left: Old Head of Kinsale still standing sentinel for coastal and transatlantic shipping.

Bottom: The lovely *Lucy*. (From *Scientific American*.)

Shoot to Kill

I experienced this morbid abnormality firsthand on a patrol during which my squad was temporarily detached from the company and sent to investigate three women wearing white *ao dais* (ankle-length dresses) who, upon the appearance of American troops, withdrew from a group of peasants harvesting rice in a far-off paddy barely visible through a break in the jungle. Taking a perpendicular route, seven of us plodded along a tree-lined trail while the rest of the company proceeded toward the day's primary objective.

After a few minutes travel we broke out into a vast open area of wet and dry paddies separated by dikes a foot or so high: a patchwork quilt of green and brown. A dozen peasants bent over at the waist stood ankle-deep in water nearby while a middle-aged man in an adjacent dry paddy tended a water buffalo. Three slender white figures shaded by straw conical hats scurried toward the trees half a mile away.

"Shoot them!" shouted Yawn, indicating the women in white by aiming his rifle and firing.

It was a long shot, but certainly within range of an M-16 in the hands of an expert marksman. Obediently I raised my rifle, jammed the plastic stock against my shoulder, and melded my eye to the sights. Taking careful aim, I squeezed off half a clip - ten bullets - and when I finally lowered my gun to look over the scene of battle, the field was clear of white fleeing figures.

The rice-harvesting peasants were crying and shouting and scampering out of the way. The man in charge of the water buffalo lay groveling on the ground, and the domestic bovine was pirouetting crazily with fear. Yawn shifted his aim to the water buffalo - about a hundred yards away - and peppered its broad brown back with slugs. At that distance he could hardly miss such a barn-sized target. The water buffalo was being stung by the projectiles, and its circular dance instantly turned into fierce frenzied leaps much like those of a bucking bronco. When the beast's owner stood up and ran after his charge, Yawn hastily shot at the man.

I trotted past a mob of screaming, scrambling women along a dike that separated two wet paddies. With each running step my web harness rose clear off my shoulders and slammed back down with a jolt. Half a dozen hand grenades crashed together on my chest with a sound like clanging bells. The weight in front shifted the harness balance forward. The bottom two grenades kept slamming into my groin so I was forced to secure them with one hand as I ran with my rifle in the other.

I didn't splash diagonally toward the man and the water buffalo because running through water impeded progress tremendously.

When I reached the intersection of a perpendicular dike I turned right and ran for all I was worth. I saw Yawn's bullets kicking up dust around the water buffalo and its owner, but so far the man, who was running around the crazed, wounded animal and was snatching for the reins, had not been hit.

When I reached Yawn's line of fire I shouted and waved my rifle over my head. He flung down his gun in disgust because my action robbed him of a kill. Yawned finished off his magazine at the water buffalo, then ceased firing. After the cacophony of barking rifles the sudden silence was deafening.

I tried to keep myself between Yawn and his target, but this was difficult to do because the man kept springing after the water buffalo's reins. My shouts finally got his attention. He jabbered in Vietnamese and gesticulated wildly at the rampaging water buffalo, and continued to sidle after it despite vociferous threats. He didn't appear to be intimidated by me or by the squad of American soldiers.

I shouted and pointed toward Yawn. "Don't run away because that man wants an excuse to kill you!"

We stood as close to face-to-face as we could get considering the difference in our heights - the top of his head barely reached the middle of my chest. I yelled at him and he yelled at me; I cursed in English, he cursed in Vietnamese. We continued this senseless babble for nearly a minute, neither understanding the words that the other was saying but both comprehending the meaning. When he started to amble after his water buffalo again despite my gesticulations and loud protestations, I added might to my threat.

I aimed my rifle at the ground about three inches from the peasant's foot. When the gun went off he leaped like a cat at least two feet into the air, and when he came down he had a different opinion about who was in charge of the situation. His volubility ceased, and he meekly permitted me to march him back to where the squad had gathered around the group of crying, cowering women.

Yawn was furious. Somehow a young girl had gotten shot in the back. She lay on the ground bleeding while the older women wailed. Frenetic grilling revealed that none of the other squad members had discharged their weapons despite Yawn's order to shoot. Only he and I had fired, so any blood spilled lay on our hands alone.

Referring to the fleeing women in white, I admitted, "And I didn't shoot *at* them."

I soon realized that my plan could never have worked. I reasoned that Yawn's order to shoot down the three women in white was predicated upon the fact that they were running away, and that to save them I had only to bring them to a halt. I laid down a precise pattern of fire directly in their path, my bullets kicking up dust and clods of dirt whose message was intended to demonstrate the futility of escape. I could just as easily have killed them in a trice. They fled all the faster and vanished in the foliage. Yawn's rounds never got close.

When I explained this to Yawn he became apoplectic. "When I say shoot, I mean shoot to kill!" he screamed.

"But the only reason they ran was because we were chasing them and shooting at them."

He was not impressed by my logic. Now that the shooting was over the women, crouching on the ground, gathered around the wounded girl and waited silently, almost expectantly, for a similar fate.

"Let's get out of here," Yawn said. "And bring those two men along."

In addition to the man I had dragged away from his water buffalo was another who had been harvesting rice with the women. Both possessed identification cards and should have been left alone, but Yawn wanted to turn them in for questioning. Yawn marched off along our return path without a backward glance.

I motioned to the wounded girl on the ground. "What about her?"

It was now obvious to everyone that Yawn had shot the girl. Whether by accident or design only he knew for certain. My bringing her condition to his attention only infuriated him more.

"Leave her alone," he scowled. "She'll die."

Winning the Hearts and Minds of the People

I crouched by the girl's side and examined the wound. The bullet had entered her back beneath the left shoulder blade, had angled upward and outward, and had exited from the back of the upper part of her arm, which hung limp. She did not act as if she were in pain. I had no dressings, not even a spare cloth or rag, and neither did anyone else, so I ripped off her shirt and tore it into lengths, then slung it around her neck so it compressed tight against the wounds and worked as a sling.

Meanwhile Yawn and I had another confrontation. I argued that we needed to have a medevac fly her to an aid station. He argued - quite reasonably - that we were deep in enemy territory, were separated from our main force, had no medic, no radio, and no way to get help.

"We'll take her with us," I said.

He still wanted to leave her there. He argued that we had no stretcher.

"I'll carry her." I lay my rifle on the ground and scooped up the girl in my arms. She didn't weigh more than eighty or ninety pounds: a slender twig of a thing with tiny bare breasts that looked like a pair of fried eggs sunny side up. She couldn't have been more than eighteen: about the same age as my wife at home.

By defiance I had committed military sacrilege.

The girl never flinched as I hefted her in my arms. I cradled her head on my left side so her wounded arm did not press against my chest but instead lay inert across her belly. Yawn shouted angrily that I couldn't leave my rifle behind. But I couldn't hold onto it because of the girl, and I couldn't sling it over my shoulder because we had not been issued straps, an artifice which ensured that our rifles were always in hand and ready for instant use. Up to this point not one of the other men had uttered a word, offered to help, or come to my defense. They were cowed by Yawn's overbearing domination.

"One of them can carry it."

No one made a move to pick it up. Yawn and I glared at each other in a wordless contest of wills: my contempt against his authority. None of the other men spoke up, and even the Vietnamese seemed to sense the explosive danger of the clash. After a long, uncomfortable moment in the hot sun, Yawn yielded. He ordered one of the men to carry my rifle.

We left the women in the paddy behind us. Yawn led the way to our point of departure from the company, which by that time was miles away on a search and destroy mission. I struggled with my burden weighing lighter in my heart but heavier on my arms. Sweat poured down my forehead and into my eyes; my throat was parched, my biceps were weakening. I didn't know how many miles I could take the strain, and Yawn did not look back to see if I was maintaining the pace.

After an indeterminate time we came upon an oasis: a two-story house in the middle of a vast open field. For a moment I thought I had drifted through *The Twilight Zone* and gone back in time to the country home of my grandfather's birth. The house sat in the middle of a large, all-around yard enclosed by a wooden fence, and a stone-walled well occupied a cool place under the shade of a large deciduous tree. Simply-dressed women eyed us warily as a host of children gamboled over the grounds. It might have been a school.

Yawn led the troops through the gate and announced that we would take a break in the shade. Gently I placed the girl on the ground near the well and gave her water. We all drank our fill and topped off canteens under the watchful eyes of the locals, who neither interfered with us nor exhibited curiosity. They simply watched. We did not exchange greetings or ask permission for anything.

I unbuttoned my shirt and let the sweat evaporate from my chest. It was deliciously cool in the shade of the tree.

I noticed an abandoned door propped up against a tree. I was ecstatic, for here was my stretcher. I appropriated it at once. When Yawn gave the order to go, I placed the door on the ground next to the girl, helped her slide onto it, then motioned for each of the two cap-

1960's: Deep-Water Triumph and Turmoil - 181

tured men to take up a corner of the foot of the door as I picked up the head with my arms behind me: like moving men hauling a piece of furniture. The procession filed out of the yard with scarcely any reaction from the inhabitants.

With renewed vigor and with much of the weight off my arms I thought I could carry my precious load until we caught up with the company, but after half a mile of walking in the hot sun my muscles ached horribly and I began to stumble. I went down on one knee, managed to pick myself up, and kept going for another quarter mile. I went down on both knees, jolting the girl horribly, and when I tried to stand I found that I didn't have the strength. By breathing deeply for a moment to recuperate my energy I finally managed to stand.

Then Yawn ordered two men to relieve me. Each one took a corner at the head of the door while I carried their rifles in addition to mine. I walked alongside in order to keep the girl in my shadow. We walked a couple more miles before catching up with the company's rear guard. A dispatch was sent forward to the commanding officer for the radio operator to call for helicopter evacuation for "two prisoners and a wounded civilian." This brought the company advance to a halt.

I knelt by the girl's side in order to keep her face protected from the sun. I propped her head in my lap, put the mouth of my canteen to her lips, and let her drink. None of the other squad members wanted to let the captured peasants drink from their canteens, so I let them have my other one. All our water was soon gone. The four of us carried on light conversation, the language barrier being partially overcome by gestures and hand signals. The girl spoke in a soft lilting voice that carried no hint of the pain she must have felt.

Little time passed before I heard the whup-whup-whup of an approaching huey. It touched down amid a storm of propeller-driven dust. I lifted the girl in my arms. A medevac team jumped down from the cargo deck hauling a stretcher. I walked toward them slowly.

The girl yanked my sleeve with her uninjured hand. I looked down and realized that she was trying to tell me something. Most of her sing-song tones were drowned out by the noise of the huey's engine and whirling blades, and I couldn't understand her language anyway, but I fully appreciated her meaning by her action. My shirt was still unbuttoned, so she leaned up and kissed me on the chest.

Then the stretcher bearers arrived. I placed her on the stretcher as gently as possible, folded her injured arm across her middle, and looked for the last time into her dark, fathomless eyes. As the stretcher bearers ran back to the huey, Yawn shoved the two prisoners past me toward the cargo deck. They climbed aboard, the huey took off, and I watched till it disappeared over the trees on the horizon. I never saw the girl again or ever heard what happened to her.

For Your Actions . . .

For the rest of the day the men in my squad avoided me as much as possible. The object of Yawn's wrath was a stigma that no one wanted to share. As we bedded down in the jungle that night, Corporal Yawn assigned me to digging a deep pit to be used as a temporary latrine. Others shared the work on a rotating basis, but I was kept in the hole until long after dark. This offered them an opportunity to commiserate with me without fear of incurring corporal punishment, and all suggested that I try to make up with Yawn because he could make life difficult for me.

Yawn lived up to their predictions. Late that night he relieved me of the dirty detail by telling me that, for my actions, he had recommended me for court-martial.

Vietnam was a crazy, fucked-up war.

Military Compassion

Vietnam was also a war that violated the very principle for which the United States claimed to be fighting. Before I was sent to a holding camp in New Jersey to await flight orders, I applied for a temporary delay in overseas assignment because my grandfather was dying and my wife was about to give birth. I went through all the official channels without finding a compassionate ear, then I went through unofficial channels such as the chaplain and the Red Cross.

Every day my mother told me how much pain my grandfather was enduring, and how throughout his conscious moments he wanted to know why I did not come to see him. I agonized over his rapid deterioration but could not obtain the weekend pass necessary to drive to the hospital in Salisbury.

A hard-nosed sergeant did his best to block the chain of command at the link that was his desk, then became irate when I waited for him to go to lunch so I could slip by to see the commanding officer. The ploy didn't work anyway because the captain couldn't care less about my personal problems. Thereafter, the sergeant made my life as miserable as possible while the Red Cross tried to buck the system in my behalf. I spent practically all day every day camped in the sergeant's office waiting for word that my leave or temporary assignment had been granted.

To cut a long and sorrowful story short, one day the sergeant called me into his office and said with a smirk on his face, "Well, private, I tried to get the old man to block this, but the Red Cross finally got your leave for you." My momentary elation fell to despair as he continued: "You're just lucky your grandfather died in time or I'd have had your ass on the next plane outa here."

I was overwhelmed with emotions: grief, sense of loss, and hatred for the sergeant's callous disregard for my feelings. I loved my grandfather more than anyone or anything in the world. I dashed out of the office before my eyes flooded with tears, and wandered aimlessly around the parade grounds avoiding the bustling soldiers and crying uncontrollably. Hours later, when I thought I could compose myself long enough to process the paperwork without breaking down, I returned. The sergeant glared at me smugly.

The sergeant was just as smug when I came back after the funeral and continued to press for "compassionate reassignment." His sole purpose in life seemed to be to obstruct the wants and needs of others. I continued my daily routine of visiting the crowded office to check on the paperwork that would delay my assignment to Vietnam until after the baby was born. The sergeant pushed just as hard to have my flight orders accelerated before the temporary reassignment could be effected. Eventually, he won.

He threw the papers at me triumphantly. "Pack your duffel, private. I got you on this afternoon's plane."

"But sergeant, my temporary reassignment should come through any day now. I have to be with my wife when the baby is born. It'll be a long time before I see them again." The tour of duty in Vietnam was one year.

"If I have anything to say about it, you ain't never gonna see that baby."

It was the final callous insult. "I'm going home till the baby is born."

"No you ain't, private. You'll pack your duffel and be back here in ten minutes or I'll have the MP's after your ass."

We both backed our words with action.

I stormed out of the office and slammed the door behind me. Ten minutes later, with my duffel on my shoulder, I was on my way to the bus station. I went home, and the sergeant sent the MP's after me.

The rest of this story is a volume in itself. Here is the short version: I eluded the MP's by escaping out the window as they were knocking on the front door of the house, then I led them on a death-defying chase through traffic until I lost them in the woods that I knew so well. My father alerted the police to be on the look-out for me. I camped in Pennypack Park where my outdoor survival skills were put to the test. During daylight hours I slept in culverts and sewer pipes, at night I roamed the streets and kept in touch with my wife and friends by phone. I managed to be at the hospital when my wife gave birth to a fine baby boy. I was the proudest father in the world. Then, after seeing that my wife and son were healthy, I had my father drive me back to the base.

I walked straight into the sergeant's office and stood boldly in front of his desk. "I told you I was going home till the baby was born.

Now that my personal problems are taken care of, I'm ready."

His reply was unkind. Instead of being squeezed aboard the next departing plane I was locked up in the stockade with violent criminals: muggers, robbers, rapists, forgers, and the like. At a kangaroo court-martial I pleaded for leniency based upon impending fatherhood. I was found guilty of desertion and sentenced to three months hard labor and loss of rank (busted from E-2 to E-1). I did not serve out the full sentence. Instead, I was sent to my original destination where I was promoted to private first class, given a rifle and unlimited ammunition, and told to act as an emissary for the United States of America and to "win the hearts and the minds of the people" of Vietnam. If you can't win their hearts and minds, kill them.

I submit that a country cannot spread freedom to other lands unless it first has freedom in its own.

My Last Firefight

When Yawn told me of the impending court-martial I was unmoved. The army, and the congress that authorized the army to act in the country's behalf, was a sham and the shame of the concept of civilization, and a disgrace to the principles of freedom on which the Constitution was founded. The "system" had gone so badly awry that its servers had lost sight of its guiding beliefs. Court-martial under the circumstances was not only an insult to humanitarianism, but it contradicted the avowed rationale for sending troops to Vietnam.

It was that deplorable military attitude that led to such atrocities as the massacre at My Lai. The army credo was, "Don't think, just do as you're told." If you can't live with yourself afterward, it's a personal problem.

When you've already been sent to hell, life can't get any worse. I was not intimidated by Yawn's exultant threat.

Events the following day saved me from court-martial, but at terrible personal cost.

On the outskirts of a hamlet that we had searched the week before lay a field of dry paddies that could be flooded from a stream through a tunnel in a tall dike. The dike stood ten feet high, with a broad flat top that doubled as a cart trail. The narrow tunnel that pierced the bottom of the dike stretched a full thirty feet through compacted dirt. As a matter of routine two men crawled into the tunnel, one from each end. Suddenly there came shouts of fear and warning. Both men crawled out backward faster than I thought possible and, once clear, each fired shots into the entranceways. In the middle of the tunnel they had found an alcove dug into one side, and in that alcove a man sat calmly in the dark. They hadn't taken the time to notice how he was armed.

1960's: Deep-Water Triumph and Turmoil

The usual procedure in such a case was to toss grenades into the hole and fragment the enemy to death, then crawl in and pull out the body parts. But this day we had an ARVN team working with us, and they wanted the man taken alive for questioning. (ARVN is the acronym for Army of the Republic of Vietnam.) I was called in with the gas mask to effect the capture.

My squad members positioned themselves around the two openings to provide cover. I donned the gas mask, crawled along the side of the dike to the streamside entrance, tossed a tear gas grenade inside, then ran behind a nearby tree. As tear gas filled the tunnel a few rounds were pumped into the opposite end in order to prevent the man from escaping in that direction and to herd him toward me.

"Di di," I yelled through the sweat-filled mask. "Di di mau." (Literally, "Run. Run quickly," in Vietnamese, but also used by American soldiers to mean "come out now.")

I poised with my rifle jammed hard against my shoulder and with my finger tight on the trigger. I expected the man to come out shooting or, possibly, tossing grenades. If he did, I was prepared to fire. Dense clouds of gas billowed out of the tunnel opening, turning the advantage toward the enemy. No longer would I first spot the man crawling out on all fours. Instead I might see grenades bouncing along the ground or hear bullets tearing through the air. I was scared.

To my relief and complete astonishment, the man announced his forthcoming by calling out in Vietnamese and walking slowly out of the tear-gas fog with his eyes open and his hands held high over his head, reiterating something in his language and shaking his head from side to side. He wore no uniform and carried no weapons on his person. I grabbed him and pulled him aside in case someone else emerged from behind. An improper seal caused my mask to leak. I was nearly blinded by tears, yet the man demonstrated no signs of discomfort, but kept up his singsong litany. He looked like a civilian in hiding who didn't want to die.

The ARVN's took the man away. As the smoke cleared enough to see, I crawled into the tunnel to look for weapons and booby traps. I found nothing but an empty alcove. When I reported my negative findings to the captain, I saw that the ARVN's had tied the prisoner's hands behind his back and had tightened a rope around his neck. They screamed obscenities at him as they yanked on the rope and slapped him across the face. I couldn't bear to watch.

The company was called to a halt for lunch while the interrogation proceeded. I ate quickly, then ambled into the hamlet to rid my pack of some of its weight. The troops in the bush were well supplied with sundries provided by the Red Cross - more than we could ever hope to consume. It was my habit to collect all the surplus sundries, stuff them in my knapsack, and distribute them to the village youngsters during patrol. Whenever I appeared with proffered gifts I found myself surrounded by little, outthrust hands. The children never smiled, but they gratefully accepted candy, toothpaste, toilet articles, and cans of rations. The women usually looked on passively.

This day I came upon two bawling children, a boy and a girl, clinging to their mother's dress. I smiled as I handed out presents under the mother's watchful eyes. She began talking in Vietnamese. I shook my head to communicate that I did not understand. She pointed in the distance to where the prisoner was being led along the edge of the jungle by violent jerks on the rope like a stubborn farm animal, then pointed to herself and to the children by her side. I replied with hand signals, and quickly interpreted her gesticulations to mean that the man was the woman's husband and that these were their children. He was another peasant caught up in the war of political domination, and who had hidden from the American troops the same as he would have hidden from Vietcong insurgents.

The lump in my throat was so large that I could barely swallow. I gave away the rest of my food. The woman accepted my tributes, then turned and walked back into the hamlet, crying, while I ran to tell the captain what I had learned. Firing broke out before I arrived, and my first thought was that the prisoner had been executed. But then I saw action on the other side of the stream that bisected the paddies. Two men flipped the lid off a spider hole and pulled out a dead guerrilla.

"Take cover!" Yawn yelled.

I saw the prisoner on the opposite bank indicate a grassy patch of dirt. More shooting broke out as M-16's on full automatic chopped thatch away from the lid of another spider hole. The prisoner knew where the VC were hiding, and he was pointing them out! Almost immediately came the rat-tat-tat of an AK-47, followed by the crescendo of responding M-16's. Grenades whumped in the ground as machine-gun bullets scythed through the trees. Shots and shouts filled the air.

"Watch out! They're everywhere!"

On the other side of the stream I saw uniformed men rising out of the ground like soldier's sown from the Hydra's teeth. Instinctively I lunged for cover with the other men in my squad, behind a foot-high dike overlooking the water through a natural bamboo barricade. The village was not simply sprinkled with isolated spider holes, but stood atop a major underground complex of intersecting tunnels. Our forces were mixed like two football teams after scrimmage. We couldn't fire into the mingling mass across the stream for fear of hitting our own men.

"They're comin' outa the river bank!"

Armed soldiers poured out of the muddy slopes at the water's edge. The fusillade that erupted was deafening. Now I had a clear shot

1960's: Deep-Water Triumph and Turmoil - 187

at the enemy. Bullets chopping through the bamboo forced me down.

"They're on your side! They're on your side!"

Either the tunnel complex extended under the stream or a separate system existed on the side where my squad happened to be.

"Look out! They're behind you!"

Popping sounds came from our rear. The entire squad leaped up from behind the dike and charged for the bamboo thicket half a dozen paces in front, temporarily forming a line. I was spinning to glance behind when I spotted a pair of NVA regulars climbing up the bank toward our left flank, no more than forty feet away. At the same time, another regular unseen by me enfiladed our rank from the right, boxing my squad in a four-way crossfire. As the two regulars to my left swung up their rifles at me, I aimed at the rearmost and burped off three rounds.

I was in the middle of a step with my left leg foremost, shifting my aim toward the other soldier as I ran, when an enemy bullet from the right hit the inside of my upper leg, tore through the femur, and splattered blood, muscle, and bone chips out the other side. The force of the blow spun me half way around. I crashed to the ground on my right side, landing on top of my rifle. The pain was instant and excruciating. "I'm hit! I'm hit!" With extreme effort I pushed myself up onto my right elbow. I tried to crawl toward cover, but my injured left leg lay across my right leg, and I felt numb and paralyzed from the waist down. I couldn't move.

My upper body pressed hard against my right arm and my elbow was pinioned, but I still retained my grip on the trigger. I cupped my left hand over the exit wound in order to stem the flow of blood.

Then I perceived that the other of the pair of NVA regulars was running in a crouch along the perpendicular dike. His uniformed partner was not in sight. The enemy soldier turned his head as he ran and looked me straight in the eyes. No prescience was required to read that look.

In sheer terror I screamed at my teammates who were digging furiously into the bamboo thicket. "Get him! Get him! He's going to shoot me!"

I twisted back just as Charlie drew to a halt. With controlled and deliberate intent the man faced me, dropped confidently to one leg, rested his elbow on his knee, and sighted down the barrel of his gun directly at my heart. He was so close - barely fifty feet away - that I could clearly see the wooden stock and protruding banana clip of the AK-47.

My terror turned to panic which I instantly restrained. I couldn't move the rifle to my cheek in order to aim it at the enemy, so I leaned back and raised my forearm barely enough to elevate the barrel off the ground. I started pulling the trigger. What is hundreds of words in the telling was only seconds in occurring. All the while I stared at the black round hole of the enemy muzzle, I looked to see where my bullets were hitting the ground in relation to my target. I didn't see any tell-tale spurts of dust, so I reasoned that I must be shooting too high.

NEAR-DEATH EXPERIENCE

I pressed my arm down flat against the ground, pulling the trigger all the while. I saw the man flinch and a muzzle flash in nearly simultaneous sequence, felt something solid and incredibly cold penetrate my chest like a sharpened icicle, formed a mental image of a hole drilled completely through my body, was slammed back against the ground so hard that I bounced forward onto my face, found my viewpoint suddenly shifted to a point inside my skull, looked down upon the convolutions of my brain, observed a thin white mist separate from the gray matter and coalesce into a cotton-like ball which shot out of my head and into the sky up through the stratosphere until I could see the entire blue and white planet hanging beneath me against the black backdrop of space. I moved faster than the speed of light past the Moon and planets and out of the solar system into the distant interstellar reaches. As I zoomed by thousands of stars every second I found myself approaching a vast sphere of brilliant white light: a living galactic core in which souls found final solace.

I felt strangely at ease. Somehow I knew that when I reached that central sphere I would be absorbed by the light, and in the process of absorption I would find freedom from pain, would find surcease from sorrow, would find everlasting tranquility, would find ultimate peace, would find - death.

Although the fear and anticipation of death can be torturous, death itself is not a terrible experience for the one who dies. It is the living and the loved ones who suffer.

But I was not ready to die. I refused to yield. By means of some mysterious mental force I willed myself to stop on the fringe of that specious sphere of light. I accelerated away from the deceptive illusion of serenity, back through space, back through the solar system, down through the atmosphere, and back inside my skull, where the ball of mist that composed my essence drifted apart, and the thin etheric wisps remaining perfused into the folds of my brain.

I awoke to a world of mortal suffering.

My right eye lay pressed against the dirt. With my left eye I could see no farther than a few inches beyond the rim of my steel pot. The rattle of rifle fire and the bursting of grenades told of the battle that still raged. Because of the nerves that were torn away by the slug, my left arm jerked spasmodically over my head as if some mad puppeteer were yanking frenetically on invisible strings. I yelled "Medic!" in a voice that barely croaked. My call was an instinctive response that sounded ludicrous under the circumstances, even to me. It was blatantly stereotypic.

I seethed with anger that my companions had dived for cover and left me to face the enemy alone. "He shot me again," I lamented.

"Gentile, lay still and maybe he'll think you're dead!" The machine gunner's stern advice comes across humorously in print, but it was not intended to be funny and in fact was good advice in light of my predicament: exposed atop a dike with my arm twitching crazily.

The fighting gradually died down. The sharp chatter of automatic weapons fire yielded to the howls and moans of the wounded. I lost all capacity to cry out, but I remained conscious. Boots thumped the ground nearby, shadows cast across my face. PFC Tye grasped my left shoulder and rolled me over onto my back. I screamed at the excruciating pain induced by the movement.

Tye unbuttoned my shirt to look at the tiny hole in my chest. He also examined my leg, which was bleeding profusely. During his ministrations I continuously asked for a medic and morphine. The best he could do was give me a sip of water, which I craved. When he said he was going to remove my web gear, I cried for him not to move me. Gently he unbuckled my equipment harness and pistol belt so I could be lifted out of the gear when the medevac arrived.

Reinforcements were landing en masse. Helicopter gunships flew so thick overhead that they darkened the sky like a horde of locusts in migration. The battle was very much in progress, but the position occupied by my squad was no longer in the arena. A colored smoke grenade marked our location for the medevac to home in on. Twenty feet away, a fellow squad member bawled like a banshee. He had been shot in the lower leg in the same enfilade as I, then later suffered a glancing shot in the chest.

A huey touched the ground in a cloud of dust. A stretcher was thrown by my side. Four men grabbed me by the arms and legs - I screamed again in pain - and lifted me onto the stretcher. The two stretcher bearers raced for the hovering huey. The bearer at my feet threw the end of the stretcher into the cargo bay, but he didn't lift quite high enough. My left heel caught the edge of the platform. The bearer at my head was already shoving me in. This forced my broken leg to bend up at a sharp angle, knee pointing skyward. I shrieked in agony despite the sucking chest wound.

The helicopter took off as soon as my companion was tossed in. The huey banked at an altitude of five hundred feet, and as my head lolled to the side I stared straight down at the battlefield, where olive-drab troops scurried over the ground like ants whose nest had been invaded. Then, mercifully, I passed out.

Last Rights

I regained consciousness in a forward aid station attached to the fire base that gave artillery support to my company. There were no doctors there. A medic wrapped an inflatable splint around my leg and secured it with Velcro, then inflated the plastic bladder from a CO_2 bottle. Fade out.

I came to again just in time to see another medic straighten and turn to someone beyond my field of vision. "You better go get a chaplain. This guy's not gonna make it."

My eyes closed and opened, and there stood a chaplain looking down at me in evident distress. I saw him make the sign of the cross on his chest. His mouth moved, but I could not hear any words. Fade out.

The sun burned into my eyes. I was being carried on a stretcher under a deep blue sky. In addition to four stretcher bearers, on each side of me a soldier scurried along holding aloft a glass bottle from which a tube ran down into my arms.

"Where are we going?" I muttered hoarsely.

"To a hospital in Qui Nhon. It's about a forty-five minute plane ride."

"Can I get a shot?" Meaning morphine.

"When you get there."

I passed out without ever seeing the plane. I came to in an emergency room just as a nurse removed the St. Christopher's medal that I wore around my neck in place of dog tags. The Catholic symbolism meant nothing to me, but the sentimental value was great because Bill Reese had given it to me on the day of my induction. I grabbed her wrist harshly. "Do you have to take it?"

"Yes," she said firmly but soothingly. "But I'll see that you get it back."

Fade out.

The pain was unbearable when the two orderlies hoisted me onto a cold metal table beneath the x-ray machine.

Fade out.

Doctors and nurses stood shoulder to shoulder completely surrounding me. Their masked faces were bathed in stark white light. Scissors cut off my fatigues and jungle boots and needles went into my arms.

"Can I have a shot now?"

A tall man wearing a white surgical gown secured a vial of clear liquid to a hanger. "That's what this is." He connected the vial to a tube coming up from my arm, then turned a spit cock. "Bye."

The anesthesia hit me fast. My head lolled back on the operating table just as a doctor lifted my left shoulder to examine the exit wound. I heard a sucking sound and saw a thick wad of bloody pulp gush from a hole in my back and splash off the table into his face. He let go of my shoulder and jerked away . . .

BACK FROM THE DEAD

My world became one of sleep, pain, darkness, pain, spectral faces, pain, shadowy shapes, pain, pain, and more pain. I suffered wakefulness from shot to shot. When my screams became too loud and long, a nurse plunged a needle into my upper arm and injected me with Demerol. This semi-conscious state went on for days - perhaps weeks. I remember snatches of events and to some extent the sequence, but have retained little sense of the lapse of time.

At first my vision extended only inches from my eyes, so that it was necessary for doctors and nurses to lean into my face when talking in order to get my attention. Days must have passed before I recognized that the indistinct form at the limit of my vision was my leg. It was suspended from a metal trellis, and eventually I learned that a stainless steel pin had been punched through the shin bone and was wired to a heavy weight at the foot of a traction bar. Because the bullet had demolished a large chunk of the femur and left a gap in the bone, the weight had been installed in order to stretch the leg apart, otherwise the muscles, acting like powerful rubber bands, would pull the shattered ends of the bone together and my leg would heal short. By keeping the leg stretched, the bone could not knit until each end grew out far enough to reach the other end. The bullet had narrowly missed the femoral artery.

A doctor informed me further about my condition. The bullet that had been intended for my heart had missed that essential organ by nearly three inches. It had entered my chest between the second and third ribs without breaking either one, passed barely an inch under the aortic arch - a pierced aorta would have proven instantly fatal - ripped through the lung, and tore out through the middle of the scapula (the shoulder blade), warping the bone and carving an exit wound the size of a silver dollar.

Nerve damage was extensive. My arm was paralyzed and totally without feeling. For weeks the doctors stuck pins into my fingers, palm, and back of the hand - without obtaining any response. If I was distracted by one doctor while another tried the procedure, I wasn't aware that they were testing me. The general consensus from medical opinions was that I would never move my arm again.

Nerve damage in my leg created a contrary condition. Intense, burning pain enveloped my foot as if it were buried in a bed of hot coals or fettered in a furnace. Phantom flames shot up my lower leg, sometimes raging out of control, at other times smoldering slowly.

Ironically, the wound itself was relatively painless. I retained partial movement of my toes. Most of my leg was numb to pin pricks, yet the burning pain was continuous.

Every breath was torture. My lungs expanded with each inhalation, which was equivalent in effect to stretching all the chest injuries, internal and external, created by the bullet's passage through my torso. Consequently, I breathed short and shallow.

Yet for all the damage done to my body I was lucky to be alive. For that I have to thank the unknown medic who, in a last-ditch attempt when there was nothing to lose - the chaplain had already conducted the ceremony for last rites - performed the cut-down that saved my life. Unable to locate a vein in my arm due to the massive loss of blood, he sliced open the flesh in the crook of my right arm, reached in with a finger and pulled out the vein - which he managed not to nick - and stabbed it with a needle to which he connected a bottle of plasma.

This timely transfusion brought me back from beyond the edge of death.

I am told that field medics receive only an eight week course in emergency first aid. A medic's prime function is to dispense salt tablets and anti-malaria pills, and, in the extreme, slap a compress on an open wound. They are not trained to perform delicate cut-downs.

How close had I truly been to the oblivion beyond? No one can say. Had my heart stopped beating, stunned into paralysis by the trauma of the enemy bullet's close passage? I don't know. Did I see god? Definitely not. My trip through space to the white galactic core was an hallucination induced by the lack of oxygen and blood sugar in the brain, and by the release of organic chemicals such as adrenaline and endorphins - all of which affect the brain and, by extension, the mind, which is a construct of the brain.

Given a similar stimulus almost any human being would see white light. Studies of near-death experience (NDE) and consequential out-of-body experience (OOB) have established that fact. The reason that some people see their creator is due to their preconceived notions and religious convictions - they see in that final moment what they hope to see, or what their cultural heritage has led them to expect to see.

My near-death experience did not convert me to spirituality. My acceptance of reality remains unchanged.

A World of Pain

As my sphere of awareness expanded I noticed a profusion of plastic tubing that surrounded me like a web. Two tubes snaked down from hanging bottles and fed blood and liquid into my arms intravenously. One drainage tube was sewn into the outer gash on my leg, another into the huge hole in my shoulder blade. They siphoned pus from the wounds. To drain fluid from my punctured and deflated lung, a thin plastic tube had been forced through the side wall of my chest and pleural cavity. A catheter drained my bladder. This mass of tubing compelled doctors and nurses to approach my bed from the right, where they had to contend with only one aerial tube during examinations. Changing sheets was a complicated job as IV bottles and drainage bags were shuffled around and the coils of tubing were rerouted.

The dim, distant shadows slowly resolved into recognizable objects. For a long time, doctors and nurses appeared abruptly in my field of vision like off-camera effects. After a while I was able to distinguish the nursing station from the background gloom. As the most critical patient in the ward my bed was positioned right next to it. Later, I could see all the way across the aisle. The patient in the bed opposite mine had suffered severe shrapnel wounds from incoming artillery. The flesh had been stripped from one whole side of his torso, revealing an ugly mass of raw tissue, the white bones of the rib cage, and exposed organs. He screamed louder than I only because both his lungs were intact.

Eventually I learned that the hospital consisted of a series of interconnected Quonset huts, whose corrugated metal walls held the tropical heat at bay as long as the air conditioners ran continuously. A war zone hospital is a place of unspeakable horrors.

One day a Vietcong guerrilla was rolled into the ward. He had been napalmed, and every bit of skin had been burned off his body. He was wrapped in white bandages from head to toe, including his face, except for slits for his mouth and nostrils. He reminded me of Boris Karloff as *The Mummy*. He screamed constantly at the top of his lungs for hours and hours on end. During that time I forgot my own pain because I couldn't imagine what exquisite torment could make a person cry so pitifully. I was deeply disturbed by his agony, and just as deeply relieved when he died. I saw his body being wheeled out late that night. In the sudden shocking silence I felt relief that such ineffable suffering had ceased. Death was a desired pardon.

Two officers appeared at my bedside wearing dress uniforms, the first I had seen since arriving in-country. They explained that they had visited twice but that both times I had been asleep. With little pomp, one read official orders and the other presented me with a Purple Heart. They stayed less than five minutes. Judging by the stack of medals they were carrying in a cardboard box, a busy day awaited them.

The army had just paid me off for my wounds with a cloth ribbon and a lead medallion. I was barely conscious enough to appreciate the gesture.

Then came more surgery. I don't know what they did to me other than to remove the catheter, but I was under the knife for four and a half hours - shorter than the initial surgery. Afterward, I was prepared for transportation by being encased in plaster from just below the armpits: a full body cast that immobilized both legs and left only my toes exposed. The traction pin was held in place by the plaster wraps. A metal spanner between my thighs maintained the angle of spread and doubled as a carrying yoke. Two men could lift me like a packing case, each with one hand on the yoke between my legs and the other hand gripping the chest cavity.

The Red Cross gave me some toilet articles and a ditty bag to carry them in. The bag was also big enough to hold my wallet and St. Christopher's medal - the only personal items I possessed - and the Purple Heart.

I was wheeled out of the Quonset hut into the hot, bright sun. My only covering beside the plaster cast was a sheet. A truck drove me to the airport where I was loaded onto a C-130 cargo jet that had been specially converted to carry wounded soldiers. Non-ambulatory patients on stretchers were stacked in tiers like bunks on a troopship. The restricted head room didn't bother me because I could neither rise up nor roll over.

Early in the flight my pain medication wore off. In the ward I was given injections of Demerol every four hours, day and night. I needed the shots desperately and on time because the pain-relieving effect always wore off before the next dose was due. I suffered horribly as the end of the fourth hour approached. A nurse was in attendance on the plane, but no doctor. She explained that she was not authorized to give prescription medicine. The most she could offer was aspirin. This was absurd in light of the seriousness of my condition. I had undergone extensive surgery only the day before.

The rest of the flight was a nightmare. And the nightmare grew worse when we landed in the Philippines, a stopover on the way to Tokyo. During my conscious moments I was wracked with pain. An ambulance took me to a holding ward for the night. I kept asking for a doctor to prescribe pain medication, but it took hours for the only one on duty to wade through scores of new arrivals and to read all their files. When a doctor finally reached me, he explained that my medical file contained no orders for pain medication, and that I would have to wait until my final destination before having my file reviewed. I groaned at this information.

His only concern was that the jostling of travel may have dislodged the set of the femur. He sent me to another building for x-rays. I waited for hours for the technicians to get to me, by which time it was long after dark. Then I was wheeled back to the holding ward, which was the size of a gymnasium but with a low ceiling. Here the non-ambulatory cases were left for the night on wheeled stretchers. I had not had any food since morning, but I was too sick with pain to eat. The room stank with the odor of unchanged drainage bags, and was filled with the moans and the cries of patients who, like me, were given no medication to alleviate their pain. To add to the overall agony, the lights were kept on all night.

I did not sleep at all but spent the entire night writhing in pain. A towel draped over my head kept the light out of my eyes. The next day I was put on a plane for Tokyo. I remember nothing about the flight except chronic pain. After landing, a huey transported me across Tokyo and gave me a grand aerial view of Japan's capital metropolis. Mount Fuji towered like a painted backdrop behind a cramped, crowded city that was stereotypically Japanese in design and construction - the narrow streets and conical houses were identical to those trampled by the prehistoric Godzilla.

The U.S. army hospital in Tokyo was a modern, multi-story steel-and-concrete building that offered medical treatment and surgical procedures not available in Vietnam. If a soldier could be cured or healed in less than two months, he was returned to duty. If, in the opinion of a board of doctors, a patient required more than two months for complete recovery, he was rotated home permanently.

I was placed in an overcrowded ward whose medical and management staff were totally dedicated to the comfort of their patients. During my stay I received the best care that it was possible for them to provide. When I arrived in late afternoon it was obvious how much I was suffering. No doctor had yet been assigned to my case so injections were not allowed, but staff orderlies obtained special approval for the administration of codeine in tablet form.

Codeine relieved my pain, but it also induced hallucinations. Late that night I came out of my stupor in the middle of a jungle scene. A firefight was in progress. As our position was about to be overrun, the officer in charge gave the order to move out. I was unable to walk. I bellowed, "Don't leave me! Don't leave me!" When no one paid any attention to my plea, I struggled after them, screaming.

An orderly rushed into the ward and found me standing beside my bed. I had managed to shove myself over the side where by great good luck I had hit the linoleum floor with both feet of the cast. Instead of toppling to one side, I remained upright like an ancient clay statue, leaning back against the tall frame. Single handed, the orderly levered me back onto the bed, then raised the side rails, which had been left down in the belief that a person in a full body cast and with one arm paralyzed could not get out of bed.

The tubes that were sewn into my body had been held in place by the plaster, but the IV and drainage bags had torn loose, creating a mess that the orderly had to put back together. Although he assured

196 – The *Lusitania* Controversies

me that I was safe in the hospital, my eyes perceived things differently. Gradually, over what seem like an hour or so, the battle scene transformed. The sandbag bunkers melted into beds, the palm trees resolved into concrete columns, the men in the squad became sleeping patients, none of whom had been roused by the tumult.

The next day I was wheeled into the preparation room to have the cast removed and additional surgery performed. Prior cases took up too much time and they didn't get to me. I was alarmed because until I was thoroughly examined by a doctor I couldn't get shots for pain. The nurses and orderlies pressed my case until a doctor took it upon himself to authorize the injection of morphine. I passed a tolerable night - without hallucinations - and had my surgery the following afternoon.

When I regained consciousness I found myself restrained in a new traction assembly that kept my broken leg angled nearly forty-five degrees to the side. The torsion was intended to help the bone heal straight. A new traction pin had been inserted through the shin bone. This one was drilled instead of punched, and was positioned deeper toward the middle of the bone for greater strength. Heavier weights beneath the pulley stretched the muscles farther apart. All the drainage tubes had been removed and the holes stitched. A single IV dribbled saline solution into a vein.

During the days and weeks that followed I was preoccupied by pain. My only alleviation came from injections of morphine and Demerol. I was switched continually from one to the other so my body would not become dependent upon the drugs. I avoided addiction, but my body produced a tolerance to the drugs such that relief no longer lasted four hours from one shot to the next.

Once injected, a delicious numbness engulfed my feet and surged up my legs and body until a cloudlike, insensate wave swept over my head, at which point I drifted peacefully to sleep. This feeling was not a "high" in the junkie's sense of the term, but a sense of well-being that resulted from the release from intolerable pain. The effect perhaps is similar. The difference lies in the starting point.

Increasing waves of pain racked my leg and shoulder as the analgesic wore off. Suffering miserably, I watched the clock as the seconds ticked by with agonizing reluctance toward the time appointed for the next hypodermic injection, which brought diminishing relief. The period and potency of sedation dropped with each succeeding shot, until finally I was fully conscious and in torment more than half the time. Instead of increasing the dosage or decreasing the time between shots, the doctors decided to wean me from the drugs by permitting injections only at night. During the day I was reduced to mild medication in pill form.

Along with this increased sensitivity to pain I found certain positions uncomfortable, as if a bed spring were digging into my backside. This forced me to lie more over to my right, where I was soon covered with bed sores. When I complained about this newfound agony, a doctor enlightened me about the wound in my buttock. This came as a surprise to me. He assumed that it was a shrapnel wound, but I knew that no grenades had gone off in my vicinity. I think that as I lay on the ground with my arm jerking madly over my head, the VC guerrilla hastily plunked another round into me and that it passed through the meat of the gluteus.

At the height of my medicinal intake I swallowed as many as twenty-nine pills a day, of various prescriptions. I had received so many injections - some analgesics but mostly antibiotics - that both upper arms became callused to the point that needles could no longer penetrate. They bounced off even in the hands of highly skilled nurses. I began receiving injections in the leg.

Once my temperature got so high that my body was packed in ice to reduce the fever.

After what seemed like eternity to me, but which was more in the order of a month or so, the chronic pain became tolerable and my level of external awareness increased to the point at which I could hold a conversation, listen to music on the radio, watch Japanese television, and read a book. I recognized my surroundings and knew what was going on in the ward, which was my only frame of reference. Most of my solace I found in the pages of books that were circulated by a Red Cross nurse, whose "bookmobile" made the rounds several times a week.

Thus I remember the buzz of activity that occurred when a group of people entered the room wearing neither uniforms nor hospital garb. Since the head of my bed faced the hallway door I had to twist around in the traction harness to see what was causing the ruckus. If I had been given a choice of celebrities I would most like to see in person, without hesitation I would have picked the very man who entered the ward unannounced. He was my teenage idol, Danny Kaye. With an entourage of Japanese dignitaries, reporters, and photographers, the famous movie and television star stopped and chatted with every patient. The photographer took a Polaroid picture which Danny Kaye autographed on the spot.

When he reached my bed he smiled and shook the hand that still functioned. When he asked me how I was doing, I lied - I told him I was okay. He offered words of encouragement. Then he saw my toe cap and his creative talent emerged. I lay naked except for a sheet and, when I was cold, a couple of blankets. Because the traction assembly prevented my leg and foot from being covered, and due to restricted arterial circulation due to the elevated posture, my toes were always cold. A Red Cross nurse had knitted a red woolen cap with a tiny tassel on top, which she planted on my toes for warmth. It was my only piece of apparel. Now Danny Kaye took a felt-tipped pen and drew a happy face beneath the cap. My skin was hypersensitive from damaged nerves so that each stroke of the pen felt like the scratch of a knife, yet I never winced or complained. I was happy to have him draw on my foot not only because of who he was but because of what he represented.

Someone back home cared about us. And, as I later discovered, there weren't too many of them.

I received another visit from an unexpected quarter, this one from Corporal Yawn. While in Tokyo on R & R he came to the hospital to see me and the other man who had been wounded in the crossfire. He was perfectly affable and exhibited no signs of animosity over our past differences. He filled me in on the details of the firefight. He also wished me luck in recovering from my wounds. I have never understood why he bothered. I didn't think he was capable of feeling guilt.

After my wounded companion recovered enough to become ambulatory, he rolled his wheelchair in to see me. (I have forgotten his name.) We compared notes of that awful day in the battlefield. The bullet that took him down in the first fusillade went through his calf between the two bones without breaking either one. Then the VC marksman who shot me through the chest tried to do the same for him. His bullet hit my friend directly in line with the heart, but he was saved from death by an extraordinary combination of circumstances: the bullet struck at an oblique angle and ricocheted off his dog tags. The force of the bullet crushed his sternum and fractured the surrounding rib cage, but he lived to tell the tale.

I could almost imagine that incredible guerrilla picking off toy ducks in a shooting gallery. According to Yawn, he was finally mowed down by sheer American firepower. I have often wondered why he chose to remain in the open and fight when he could have run for cover from an overpowering force. Was he deranged or patriotic? Or do both words mean the same?

Slow Recovery

I had other memorable moments in the hospital. One day the glass on my night stand chattered. Soon the water in it was splashing and the glass vibrated across the table top. Then the windows rattled, dishes and glassware fell and broke, my bed danced over the smooth shiny floor, IV racks toppled, pandemonium reigned. Someone yelled "Incoming!" Several ambulatory patients - one in a wheelchair and another missing a leg - dived for the floor and crawled underneath the nearest bed, cowering and crying in fright.

The rumbling lasted for nearly half a minute before subsiding.

Orderlies rushed into the ward to quiet the patients in alarm before they hurt themselves in their fear-crazed belief that they were back in Vietnam and under attack. The situation was under control by the time the after shocks struck. The earthquake had moved the beds several feet from their allotted positions. There was no structural damage to the building.

Then came the day that my stitches were removed. These great wire sutures were over an inch long and nearly as thick as a coat hanger, with the ends twisted together like spliced electrical cables. I wanted sodium pentathol for the painful operation but I was not even given Novocain or a local anesthetic - not so much as a sedative. A medical sergeant clipped the wire with cutters, then yanked each one out from the end with the splice, even those that were partially overgrown with skin. I was a bloody mess at the end of the procedure.

The only treatment that didn't hurt was the cleansing of the hole left by the first traction pin. Antiseptic solution squirted into one end of the hole then flowed through the bone and spilled out the other side. This was done until the skin healed over the opening. My leg slid back and forth along the traction pin in use without any feeling whatsoever.

Once I yelled for an orderly to look at what I could do. I felt my finger twitch, and by concentrating hard I was able to make it twitch again. When the doctors next made their rounds I demonstrated my progress. They smiled and nodded, poked and prodded, and suggested that although some of the damaged nerves were healing, I should not get my hopes up for total recovery. Within days I was able to flex my fingers, then make a weak fist, then drag my arm. A week later I could lift my forearm. The doctors were astounded. I was ecstatic. Now I could hold a book in both hands.

I could also hold pen and paper. Not until I received a reply to my first brief letter did I learn that the army had failed to notify my family of my injuries. No one at home knew that I had been wounded. They thought that I had simply stopped writing.

I spent a couple of months in the hospital in Tokyo without ever leaving the ward. Other patients were wheeled down the corridor and out into the sun on the balcony, but because of the angle of the traction assembly my bed wouldn't fit through the door. Finally the doctors agreed that I had recovered enough to survive the long flight to the States. This time I was conscious as the traction device was removed and a cast was molded around my body. My right leg was free from the knee down. My legs were still spread wide with a carrying yoke, but I could fit through a doorway with ease. A couple of ambulatory patients rolled me along corridors, down the elevator, across the parking lot, and to the movie theater on base. I finally got to see some recorded entertainment in which the dialogue was not spoken in

Japanese.

I left the hospital as I arrived, wearing nothing more than a sheet over the body cast and possessing no more personal items than could be stored in my Red Cross ditty bag. Only this time I was fully conscious and no longer needed pain medication to make bearable the passage of time. My arm regained some of its motion if not its strength.

My second and last view of Tokyo was seen through the window of the bus that wound through the crowded streets on the way to the airport. There I was lifted off the stretcher and carried onto a C-130 cargo plane which, like the one that brought me to Japan, was fitted to carry stretcher patients in tiers. A long time later we touched down on U.S. soil in Anchorage, where Red Cross nurses swarmed aboard the plane to cheer non-ambulatory patients with their presence and welcome-home smiles. Praise the Red Cross! That organization was always on hand to make up for what the military lacked.

From Alaska the plane flew nonstop to Andrews Air Force Base, outside Washington, DC. Our arrival in the nation's capital evoked no attention or fanfare. No bands played, no welcoming committee greeted us. We were slipped into waiting vehicles like unwanted garbage, driven to nearby Fort Meade, and checked into the hospital. I had the distinct feeling that we were being sneaked into the country to escape the notice of the press and public.

Protocol called for returning patients to be stationed in the military hospital that was closest to home and that could provide the necessary treatment for his injuries. The doctors at Fort Meade recognized immediately that I should be transferred to Valley Forge General Hospital, which specialized in orthopedics, so instead of carving me out of my cast and setting me back in traction, they placed me in a holding ward and advised the administrative staff to prepare appropriate orders.

An orderly wheeled my stretcher to a telephone so I could call home. My wife and son were living with my parents. Ironically, my father was also in the hospital, with hepatitis. My mother couldn't get away to see me, but Jay and Kenny Enright drove three hours to bring my wife to Fort Meade. My son, Michael, was with her, but regulations did not permit children on the ward, and I was not permitted to leave.

It took over a week for the army to approve my transfer. During that time I itched abominably in the sweaty cast and received no medical treatment, not even inspection of my wounds. I resided in medical limbo. After endless paperwork the army arranged to transport me by ambulance, along with two ambulatory patients and their crutches, to my final post. The drive was uneventful.

Valley Forge General Hospital was a converted two-story wood-frame structure leftover from the days of World War Two, when it served as a temporary barracks. The hospital and associated wards were not separate rooms in a single building but rather a sprawling network of previously isolated two-story wings connected by long narrow corridors that were built and adjoined later. The floor plan resembled the random lay of tiles in a completed game of dominoes. This was my home for the next ten months. Other patients lived there longer.

While traveling halfway around the world I spent a week and a half in a plaster shell which I was eager to have removed. The day after my arrival a technician using an electric buzz saw sliced through the sides of the cast. He and an orderly lifted off the carapace to expose white, pulpy flesh that couldn't have looked much different from that of a scalped turtle. Under a doctor's direction they gently lifted me out of the bottom mold and placed me in bed under a traction assembly, to which I was immediately rigged. After examining my wounds the doctor pronounced everything in order. I was dressed, as always, in a sheet.

My recuperation from this point was a long and boring process but not particularly noteworthy. Basically I had to heal. Time passed slowly.

I Become Ambulatory

Once Bill Reese brought my wife to see me. My uncles and aunts and friends of the family came to visit. Red Cross nurses, VFW members, church groups, envoys from local fellowship clubs such as the Moose and Elk, and community leagues, all paraded through the wards unannounced and without predictable schedule, often passing out snacks and gifts and always spreading cheer.

Somehow my parents could never find time to make the hour-long drive. They always seemed to be busy.

Soon I had good use of my arm. A physical therapist stopped by daily to help me exercise and build up my strength. X-ray examinations turned out well, so it wasn't too long before the doctors decided I could be released from the iron maiden. Once free from the traction assembly I was able to sit up in bed. I got so dizzy that I nearly passed out. In order to acclimatize to the altitude, I asked to have the bed cranked up in stages. A doctor slid out the traction pin without a hint of pain. The remnant hole had to be treated with antiseptic like the previous one.

After nearly four months of confinement on my back, the freedom to lie on my side was sheer luxury. The doctors warned me not to place undue stress against my left leg because the surrounding musculature was greatly atrophied and would not provide adequate support for the still-healing bone. Another patient had rolled over and rebroken his leg, then had to spend another two months in traction. I was careful.

The red-letter day was the one on which I got my very own wheelchair. Getting into it was an ordeal. A nurse and an orderly helped to pull me upright and swing my legs over the edge of the bed. At that point I was so woozy that I turned white and went limp, and had to be laid back down. My heart wasn't used to pumping blood up to my head. I fared better on the second trial, although I would have fallen flat on my face if they hadn't been holding on to me. As I slid off the bed and stood momentarily to make the transfer to the chair, I found that my good leg wasn't strong enough to support my weight. It buckled like a strand of wet spaghetti. I collapsed as they eased me into the seat. All I could do at first was sit and recoup my energy. The orderly pushed me through the corridors on a familiarization tour. There were no maps.

For days I practiced wheeling myself gently around the ward. It felt great to be mobile again, and to flex my developing muscles. At the end of a week I was strong enough to climb out of bed on my own, with the help of an overhead bar. Soon I was able to climb back in again. I was growing less dependent on the medical staff and becoming less of a burden. No longer did I have to ask for a urinal or bedpan - I could wheel myself to the latrine. Life was full of simple pleasures.

Being ambulatory after a fashion meant that I could visit the hospital library. This was not a lending library. The shelves overflowed with books that were donated by individuals, local businesses, and charitable organizations. It was a free-for-all for anyone who wanted to borrow books for keeps. I have always read prodigiously, but never so much as I did during that year I spent in the hospital. Reading is my fondest pastime.

A hospital is a depressing place. Most of the patients at Valley Forge were not going home intact. Unlike in the movies, wounds resulting from actual warfare are seldom clean and bloodless. The quick merciful death - with time for that last smoke and a soul-wrenching message for the wife and kids - is a contrivance of theatrical fiction. The teeth-gritting flesh wound is a plot device with dramatic appeal. The short, painless rehabilitation is a make-believe story meant to soothe the souls of gullible voters. In reality, bullets and shrapnel cause horrible damage and bequeath mutilations the nature of which often exceeds the capacity of the most vivid imagination. Wartime military hospitals are charnel houses of the living dead.

The hallways thrived with maimed and mangled bodies and with mindless brains. Here were not the heroes pictured in posters promoting the fight for democracy, but were hidden the permanently disfigured martyrs summoned against the threat of imprisonment to be the cat's-paws for the country's secret political ambitions. Here the norm was the disabled soldier: men missing fingers, hands, arms,

1960's: Deep-Water Triumph and Turmoil - 203

legs, eyes, groins, organs, patches of flesh, chunks of bone, pieces of head, parts of face, and some or all of the above. To truly appreciate the horrors of war, one has only to visit the wards and see what came to pass when Johnny got his gun but did not come marching home again.

Fate had a more fortunate end in store for me. While I was going to physical therapy, others were being fitted with artificial limbs or learning how to write left-handed. While I was doing wheelies down the corridor in my wheelchair, others were adjusting to severe handicaps that would leave them crippled for a lifetime. Of all the patients in Valley Forge I considered myself the luckiest, for I knew that one day, however far in the future, and no matter how much enfeebled, I would walk out of the hospital on my own two feet and with the anatomy I was born with.

Into this cheerless world came those who managed to shed some joy. Undoubtedly the one most remembered by all was Philadelphia disc jockey Jerry Blavat, known at the time as "the geeter with the heater" (whatever that meant). He arranged free entertainment for the troops and brought to the hospital such well-known singers as the Four Tops and the Temptations and other popular groups. The auditorium was always jammed on these occasions with folding seats, wheelchairs, and movable beds - standing room only was an unacceptable circumstance considering the nature of the audience.

Ambulatory patients were expected to pay for their keep, so along with the wheelchair came a job. For an hour or two each morning and afternoon I sorted letters in the mailroom. Although sitting upright for such a long period of time often caused me great discomfort, it felt good to be productive again. Having a fair amount of freedom in my schedule, I worked my job in with my physical therapy and doctor's rounds.

After a month or so in a wheelchair I slowly graduated to crutches. The process was not painful, just exhausting, especially since I had to learn how to walk again. At first I had neither strength nor sense of balance. Standing vertical was a new sensation that cannot be imagined by one who has not spent month after month on his back. An orderly held me up by the shoulders while I shuffled across the floor. During the transition stage I wheeled myself around the hospital, then walked short distances in the ward or at my destinations.

I stepped on a scale in the physical therapy room. After all the months of recuperation and building up my body, I weighed 125 pounds.

BACK TO THE REAL WORLD

Once I demonstrated to the doctors' satisfaction that I could get about on crutches without swooning or falling down, a great opportunity arose - that of obtaining an off-base pass. In order for that to happen the army first had to issue me a uniform since all my belongings were left in Vietnam due to my hasty and unforeseen departure. My duffel was never forwarded to me. This created a problem because the army had also managed to lose all my records, including my pay records. All I had was a medical file. So the army decided to *sell* me a uniform on credit until they straightened out my records and issued me a pay voucher. They also let me draw a partial pay based upon the months I had spent in the hospital.

When at last I was permitted to go home for a weekend, I couldn't find anyone to pick me up. My wife didn't drive, my parents couldn't be bothered, my friends were either working or away in college or serving in the military. No public transportation extended to Valley Forge. Fortunately, at the last minute, I met Richie Camburn, a fellow patient who not only lived in northeast Philly but who kept his car parked on the base. He offered to drive me home.

The world I came back to was not the one I had left. Even today it is difficult for me to understand the dark mood that pervaded the nation during those years known as the Vietnam era. At the time I naively expected that a lone soldier in uniform and leaning on crutches would stir some amount of sympathy, yet on subsequent passes when I hitchhiked home I found it nearly impossible to thumb a ride. I could always get to the turnpike entrance by asking around on the base, but then I might stand at the toll booth for hours before a kindly soul would stop for me.

The epitome occurred during the week of Christmas. Snow fell heavily from a nighttime sky and swirled frigidly around my feet. The thin army dress shoes offered scant protection from the cold of the winter storm. I leaned heavily on my crutches at a busy intersection only a mile from home, where a Samaritan had dropped me off. As the cars stopped for the traffic light I held out my thumb, clearly visible in the yellow glow of overhead incandescents.

People saw me. They even stared at me. They looked right into my eyes. And I looked right back into theirs. Their faces were blank and expressionless but their eyes spoke volumes. I saw hatred in those eyes, and loathing. These people even turned their heads to continue staring when the traffic began to move, leaving no doubt about the way they felt. Mine was a fleeting glimpse into the dead soul of the American spirit.

After a couple of hours, shivering and half frozen, with snow accumulating on my shoulders, and feeling lonely, dejected, and totally alienated from my fellow man, I hobbled on my way. This was only the beginning, the mere tip of the iceberg, of the treatment I later received from a civilian populace who were more crippled than I.

It was America's lowest ebb.

More Surgery

My medical problems were far from over. Months passed, my arm grew stronger, I graduated from crutches to a cane, and I should have been well along the road to recovery and discharge from the army. But the more weight I put on my wounded leg the more my foot began to ache. My foot was still hypersensitive from nerve damage, and since this was where the pain originated I made the obvious correlation. The doctor thought differently. He observed my gait and detected a limp that appeared unconnected with the gunshot wound and with the fact that despite the heavy traction weight my femur had healed a little short.

X-rays proved him right. The ankle ligaments were broken and no longer supported the foot. The damage was sustained when I jumped out of the way of that dud grenade. Now, after ten months in the hospital, I needed an operation to stabilize the ankle. The surgical procedure was neither long nor complicated. A tendon was transplanted from my lower leg and threaded through holes drilled in three adjoining bones, then tied and snugged up tight. The result was a permanently stiffened ankle. To heal properly, my lower leg was placed in a non-walking cast for two months. Then I spent a month wearing a steel brace that was built into my shoe.

By then I had spent thirteen months in the hospital, had already served beyond my two-year enlistment, had cost the government an incredible amount of money, and was owed more than a year's back pay. The army wanted to get rid of me, society wanted to forget me. I was given an honorable discharge and an application for disability pension. I was a civilian again.

Battle Syndrome Retrospect

Veterans of the Vietnam conflict have been accused of being reticent and uncommunicative about their experiences in the war zone. Psychologists have postulated that this silent circumspection might be a modern form of shell shock or the result of battle fatigue. But the general populace believes that veterans are ashamed of their military conduct, that they are ridden with guilt over what they did in the field of battle, that their consciences are so distressed by their vile and heinous actions that in order to avoid making embarrassing admissions of deed they have refused to talk at all about the conflict or even acknowledge participation. The supposition is that the dark iniquitous secrets and hateful memories have been intentionally repressed because, in light of their reprehensible behavior, that is the only way that they can live with themselves.

This response to a situation of stress - a form of post-traumatic stress syndrome - has been called the "Vietnam syndrome."

Vietnam vets have been harshly and unjustly censured. The truth is not that veterans were unwilling to share their wartime experiences with the public, but that the people were not willing to listen.

Whereas veterans returning from previous campaigns were feted as many-splendored heroes, Vietnam vets were treated as trash, scum, and baby killers. Veterans were forced to maintain silence in order to avoid persecution. I experienced this pathology firsthand. People became distant when they learned of my veteran status, then wanted nothing more to do with me. I was hounded out of social gatherings. I was openly accosted. I was shunned.

The only way to be accepted by society was not to mention my veteran status. So I quit talking about what people did not want to hear.

What created strife and bitterness among Vietnam vets was not their Vietnam experience, but their subsequent American experience: the "America syndrome." The American people misplaced the hostility they felt for their elected leaders - those who were responsible for initiating and escalating the war - and castigated the soldiers because they were more reachable.

In many ways, living as an outcast in a homeland full of odious ingrates was more stressful than being constantly under fire.

The War in Perspective

As I look back on my army episode I realize that I was a failure as a soldier. My commanding officer in basic training branded me as having "a poor attitude toward military service." I can't disagree with him, but I take exception to the implication that such an observation is categorical with respect to the full measure of life. That I react with strong opposition to threats, humiliation, and unreasonable demands, I willingly admit. "Theirs is not to reason why . . ." is a sentiment I can never embrace.

On the contrary, a good soldier must take orders mindlessly and without fear of remorse over the consequences of his actions. He must not exercise emotion or free will. He must be an automaton, an instrument, a tool that is wielded by a superior officer the way a hammer is wielded by an ironsmith. Wars are not won by those who challenge orders, but by those who carry them out, however ruthlessly.

War is anonymous and impersonal.

War is a condition in which immorality has been legitimated.

There is no room in war for charity. You fight to vanquish the enemy - or there's no sense in fighting. Halfway measures don't work. I confess that I lacked the resolve to crush an invisible enemy through the attrition of the civilian population that was forced by threat to support it. But perhaps my greater blindness lay in my inability to recognize that not all the combatants in Vietnam carried weapons, that the enemy was ubiquitous: in every village and hamlet, in every woman and child, in all the hearts and minds.

Corporal Yawn was a good soldier. My platoon sergeant was a good soldier. Each was a successful product of army indoctrination as prescribed by American acculturation. Each displayed the strength that I lacked and that was needed to achieve political conquest through military aggression. They carried out orders with unquestioning vigor and determination unbothered by guilt. They accepted unflinchingly the martial creed that the rules of conduct in war necessarily oppose the rules of conduct in peace, that war suspends all laws of humanity, and that the end justifies the means. Their behavior in combat was the only way to achieve victory against an entrenched, implacable foe.

Because I couldn't accept such tenets on blind faith, I never developed the resolve to proceed as an effective fighting machine. I was guilty of being misled by an innate sense of righteousness. I saw innocence when I should have seen hostility. I saw good in a background of bad. But my greatest weakness as a combat unit was the belief that I had the right to make moral judgments in a war that I did not understand.

Despite this admission, I suffer no guilt over my conduct in battle. Due to the kind of firefights in which I was engaged, there was too much confusion for kills to be confirmed. When armed soldiers shot at me, I returned fire with a firm intent to kill. Call it self-defense or the will to survive or any other self-serving platitude that fits. If by some chance I failed to take out my target, it was not for lack of trying.

An army of amateur soldiers who think as I did could never win a war like the one that was fought in Vietnam, which found wanting a firm objective beyond pacification by force. This is not to say that amateur soldiers can not be molded into an effective fighting force, only that they need conviction and a justification for their actions that is more powerful than the threat of criminal discipline. National defense is a far greater calling.

Although I saw the war through myopic eyes, the passage of time has granted me greater vision. I can now look back on events from the perspective of a blind, telepathic being. I can no longer see the uniforms of the men intermingled in battle, but I can read the minds of the opponents, which I find indistinguishable. None are fighting for land, for gain, or for conquest. They are fighting for what they believe in: their perception of freedom. Who is to say that when two of us met on the field of battle, one was right and the other was wrong?

Part 5
1970's: Decompression Comes of Age

A Cut Above – or Below

In 1970, the Eastern Divers Association was practically unknown outside its own membership. The trips that Elliot Subervi arranged for the club's elite divers were undistinguished, spiked only by an occasional trip to the Mud Hole. Wreck destinations rarely exceeded 130 feet. What separated EDA from commonplace dive clubs then in prevalence was not the depth of its scheduled dives but the expertise of its members. At a time when the majority of people dived with single tanks, EDA members never wore less than doubles and many carried pony bottles. When most people made dives of short duration well within no-decompression limits, EDA members planned decompression stops that put them at increased risk. When conventional divers were content to observe the exterior of intact wrecks, EDA members made deep penetrations into black foreboding interiors.

Performance that was considered excessive for "recreational" divers was the norm for EDA. This divergence of skill created a schism between the "conservatives" who indulged in wreck-diving as a mildly interesting pastime that required a minimum of training and a modicum of care and acuity, and the "extremists" who were highly motivated to excel above the ranks by diving deeper and staying down longer than was prescribed by the certifying agencies that established the standard requirements, which in turn influenced the attitudes that broadened the gap between viewpoints.

The extremists looked down on the conservatives as minimally skilled "tourist" divers, while the conservatives glanced aside at the extremists as hotshot "crazies." Thus was born the concept of the "gorilla" diver: a brawny throwback with anthropoid characteristics and a simian brain to match. An unbiased outlook on both camps would portray "tourist divers" as those who were content to spear a few fish, catch a few lobsters, pick up an easy souvenir in passing, and enjoy the social amenities offered by the club, all while practicing extreme safety, but who clearly expressed a more adventurous nature than those whose entertainment consisted of playing cards, watching television, or going on amusement park rides; and "gorilla" divers as those who aggressively met the challenges of depth, duration, and decompression not necessarily to satisfy deep-seated inner drives in the pursuit of personal peril, but who looked upon such challenges merely as barriers to overcome in the achievement of goals that were inherent to the individual but meaningless to others who harbor different ambitions.

No degree of involvement or point of view is more valid than any other. No diver needs to justify an activity that suits his needs. And scorning differing aspirations is an exercise in conceit.

A New EDA

EDA would not have achieved its potential had it not been for a combination of circumstances that occurred when the club was at its lowest ebb. George Hoffman was an industrious worker who saw where money was to be made, and set his sights on making it while paying for his diving to boot. He became a licensed charter boat captain and graduated from a Boston whaler to an aged wooden clunker that could carry six divers uncomfortably. He called his boat the *Sea Lion*. He did not give up his well-paying job as an elevator constructor. He worked during the week at his primary occupation, then spent the weekends running a dive boat: a grind he maintained for the next two decades. Since he was a member of EDA, Subervi threw occasional business his way. Hoffman kept the *Sea Lion* at Manasquan Inlet, a location that offered access to many favorite wrecks and that was only a short ride to the Mud Hole.

Subervi tired of diving and began to pursue other interests. His involvement in EDA waned. He continued to charter boats and schedule destinations, but often did not show up for the dives. When he had other things to do, he appointed someone else to take charge of the trip (that is, to collect the money and pay the captain), for which the appointee got to dive for free. This incentive appealed to Tom Roach, a telephone lineman, as a way to support his growing underwater habit.

Roach proved to be a fortuitous choice as Subervi's successor in leadership. It is no exaggeration to state that he fired EDA with new drive and enthusiasm. The organization languished during the transitory phase, but when Subervi finally stepped aside and let Roach take complete control, the latter's personality dominated the club and led it charging into deeper, darker waters.

I was fortunate enough to be at the right place in time to meet Tom Roach during EDA's transitory phase when he needed fresh bodies to meet the club's failing charter obligations.

Restarting Life

By then I had regained some of my physical fitness. This was partly due to the body's natural healing process. But not to be discounted was an active occupation, my penchant for hard work, and a strong desire to overcome my disabilities. After my discharge from the army I could not afford the luxury of college. I had a family and financial obligations to meet. I joined the union and took the required four-year apprenticeship to become an electrician in the commercial construction industry. Climbing ladders and scaffolds, scaling pipelines in refineries, walking high steel, pulling cables, and carrying heavy conduits, all contributed to rebuilding my strength. I walked fast and took steps two at a time. Whenever possible, I used my left hand and arm instead of my right - painful though it was.

My back pay from the army went toward the down payment on a house, but my starting wages barely sufficed to cover the mortgage and utility bills and keep food on the table. I settled down to lead the life of *Father Knows Best* that I grew up watching on television. I also set out to fulfill some of my childhood dreams. The one that took precedence was the exploration of caves.

Sewers were a poor substitute for caves, but as an adolescent without means of travel, the concrete storm drains offered the only means to satisfy my urge to pursue unknown regions underground. Now with a car and occasional weekends off, I actively sought the grottoes that I read about during my teenage years. The few commercial caves within a couple hours drive from home provided fun family outings, but the colored lights and guided tours were not to my liking. I wanted to see caves as they existed in nature and I wanted to discover for myself their labyrinthine passageways.

My personal library included several volumes from various state geological bureaus which not only described those caves that had been surveyed, but gave coordinates and directions on how to find the entrances. I purchased the appropriate topographical quadrangle maps, then plotted the caves on paper. Now all I needed were companions to share the adventure. Of the several friends who attended my subterranean excursions, the one most steadfast was my childhood friend, Tom Gmitter.

Caving Adventures

These pilgrimages to the cave belt in western Pennsylvania, Maryland, Virginia, and West Virginia were exciting and wrought with adventure. I garbed myself in old clothes that were soon covered with mud, wore a hard-hat to which a carbide lamp was secured, carried flashlights, candles, matches, ropes, map and compass, food and water, and spent hour after hour crawling through narrow, twisted corridors and climbing over abysses in vaulted rooms. Spelunking, as the activity is called, was as challenging as I had imagined and every bit as fulfilling.

The most stimulating allure was not in following the lines on a map to where other spelunkers had gone, but in finding new passageways that were previously unexplored. I went to great lengths to extend my explorations beyond the boundaries drawn in the books. Once I noticed a tiny notch at the base of a muddy flowstone wall which the map showed to be the end of a room without other egress. When I stooped to shine a light into the narrow crevice I saw that the opening expanded slightly before terminating suspiciously. The crevice was difficult to negotiate for several reasons: its height was less than the beam of my hard-hat, its width was only slightly broader than my shoulders, and it looped down and up like a roller coaster track.

I lay flat on my belly and squirmed forward until my hard-hat jammed in the rock. By removing the hat and holding it upright I was able to push it forward in front of me. I couldn't crawl with my arms underneath or by my side. Instead I had to stretch them out in front and pull with my fingers while pushing with my toes. I turned my head sideways and slipped through the narrowest section, but my chest was too large to fit. The only way to squeeze through was to let all the air out of my lungs, shove forward quickly, and inhale on the other side of the restriction. Then only by bending my back sharply backward could I curl past the upright barrier into the chamber beyond. No one followed me. After my solo exploration I found that I could not bend my body properly to crawl back out. I was forced to exit feet first. When I reached the restriction I exhaled completely as my friends pulled me out by the ankles.

Of all the hidden grottoes I explored, however, Laurel Creek Cave in West Virginia was by far the largest and most memorable. The entrance was 110 feet wide and 40 feet high in the middle. The tunnel extended at those same dimensions for half a mile into the side of a hill, from where a narrower tube led to a chamber that was so high and wide that our lights couldn't reach the opposite walls. Only by stationing ourselves at various locations around the perimeter of rock could we appreciate the true dimensions of this vast, vaulting cavity: the underground amphitheater was hundreds of feet across. We never saw the roof.

We spent three days exploring the many galleries and offshoots, and camped inside like troglodytes when exhaustion overtook our enthusiasm. Drinking water was plentiful because of the eponymous creek that flowed through the lower reaches. We ate out of cans and cooked on propane stoves. Our clothes and sleeping bags absorbed moisture from the constant high humidity, while drops of water that condensed on the ceiling fell like rain. This subterranean existence obviated the need for watches. We relied strictly on our bodies to tell us when to eat and sleep.

According to the book, ground-water studies indicated that Laurel Creek Cave was connected to another cave system whose resurgence through a sink hole occurred several miles upstream, but the connection had never been established. The map showed an unsurveyed tunnel which we found with ease. By poking around the boulders and tiny side passages, we picked up flowing water above its appearance in the main complex. From there on we found ourselves in unexplored territory.

The corridor was as big as a stone castle hallway. The stream stretched from side to side. At first we tried to keep our feet dry by straddling or by clinging to finger ledges. After a while this method of travel became so difficult, required so much energy, and slowed our progress to such an extent, that we abandoned it for wading. The knee-deep stream occasionally rose to our waists, at which times we hoisted our packs high on our shoulders. The water was cold but not uncomfortable. The bottom was covered with granular sand much like a desert creek bed. The water was crystal clear, revealing fish and crayfish that had adapted to the permanent nocturnal existence. Insects a-plenty provided them with food, and were also preyed upon by spiders and monstrous thousand-leggers that skittered along the walls. Laurel Creek Cave was a self-sustaining microcosmic environment.

We plowed along this gargantuan corridor for six solid hours. I was getting hungry and tired, and I dreaded retracing our water-filled path to our camp near the mouth of the cave. Then I heard a plop in the water at my feet. When I directed my light at the concentric rings that spread from the point of the splash, I saw a frog kicking mightily away. Crayfish survive in the absence of light, but I had never heard of frogs living deep under the earth. We must be near an opening to the surface. But was the entrance big enough only for frogs, or could people also pass?

We shaded our lamps, and by squinting and peering ahead we imagined a faint glow ahead in the darkness. Toward this glimmer of light we sped, hoping to find an exit. We did. Sun rays pierced down from high overhead, illuminating a ghastly sight: the opening was filled with balls of barbed wire and on the bottom lay the skeleton of a cow. I was determined not to retread the waterway, so I clambered over the decaying corpse and tugged at the mass of wire. It took a while, but we cut, tore, and fought our way through the sharp-pointed barrier, then used the metal strands as rungs to climb out of the hole. We found ourselves atop a hill in the middle of a pasture filled with grazing cattle. The rancher had piled rolls of barbed wire around the sink to prevent further loss of his herd.

Rolling, green, grass-covered hills extended all around us. I pulled out my map and compass. I had no idea where we were, but I knew the direction in which we had traveled. The cows did not seem to mind our presence in their midst. We descended the hill to a wooden fence which we climbed to gain the road. It took over an hour of zigzag hiking along dirt roads and fields to relocate the entrance to Laurel Creek Cave, where our food and equipment were stored.

Turning Point

The grotto that changed the direction of my life was Blue Hole, Virginia. Here a sink hole 90 feet deep and more than 100 feet across dropped straight down to an underground river that was thirty or forty feet wide. My two companions and I climbed down to a ledge about ten feet above the water, where we pumped up a four-person inflatable raft. We lowered the raft down the vertical embankment to the startlingly lucid surface. Then we climbed down a rope one by one and took our places on the bulging pontoons.

The downstream direction was blocked by a submerged ceiling, and the underwater passageway was choked with trees, brush, and other debris that had fallen into the sink from above. We paddled upstream for a couple hundred feet to where a muddy corridor emerged at the level of the water. We tied up the raft, then disembarked and explored the adjacent rooms and corridors. Stalactites and stalagmites adorned some of the walls in the chambers. Later, we continued upstream in the raft under a gradually lowering roof. Finally, the rocky overhead dipped down beneath the surface and prevented further progress.

I could see the bottom clearly under the raft, to all appearances practically within reach. I was shocked when I determined by means of a stone on a string that the depth was fifty feet. This intrigued me because the passageway went on as far as my light could shine. It just happened to be filled with water. I wondered about that submerged corridor, imagining that the ceiling must rise again into air-filled chambers like the one in which we floated, and which must inevitably open into other dry passageways. An underground river of such dimensions did not simply disappear.

This was no idle wonder on my part. I resolved to force the flooded passageway and to explore the unknown cavern beyond. The task could be accomplished only by means of scuba.

Diving Certification

Until that moment I had no interest at all in the underwater world. But I was determined to learn to dive in order to explore this and other flooded caves I had encountered. From the phone book I found that the Young Men's Christian Association offered scuba diving lessons, and that courses were held in one of northeast Philly's local chapters. I called for information, and signed up for the next available course.

My instructor was Bob Wilson, a Philadelphia police lieutenant who co-owned a dive shop and taught scuba in his spare time. Courses were not as formalized then as they are today. Wilson taught from a curriculum he developed himself, using as a text book Joe Strykowski's *Learning to Dive*. The YMCA provided a classroom for the lectures and an outdoor pool for hands-on equipment familiarization. We met one evening a week for ten weeks. Each three-hour session was equally divided between lecture and pool work.

I wish I could write that I passed the course with flying colors, but I can't. Because of my college background and my interest and extensive reading in all disciplines of science, theory and the written exam presented no obstacle, but because of my physical impairments, which I didn't think to mention to Wilson, I barely completed the half-mile swim in the required time, and almost flunked the ditch-and-don in the pool. My damaged lung severely reduced my cardiovascular capacity: I couldn't hold my breath for more than a few seconds, and I couldn't sustain high-speed exercise for more than a couple of minutes without getting out of breath.

It was during this time that I first met Roach, when he stopped by the pool on occasion to hobnob with Wilson. He barely spoke to me other than to acknowledge my presence. Mostly he talked with Wilson about diving in the ocean. Wilson always got violently seasick on a boat, so his in-water experience was limited to quarries, lakes, and swimming pools.

Wilson signed my certification card without reservation, so the summer ended with my becoming a duly certified scuba diver. This didn't do much for me as far as achieving my goal in Blue Hole. As yet I knew no divers other than my instructor, his aides, and my fellow students. I owned no equipment. And, as I was quick to comprehend, the possession of a C-card does not a diver make. The check-out dives I made in the quarry did not qualify me to dive in more challenging environments such as those I had in mind. Scuba certification meant only that the holder had passed a minimum competency exam, and was thereby licensed to have tanks filled with air and to breathe from them underwater. I needed experience.

DIVE CLUB OFFER

Not until the spring of 1971 did I have the opportunity to acquire some of that experience. That was when I received a phone call from one of Wilson's aides, Helen Link, who worked for the YMCA as a swimming instructor and in related capacities. A new dive club was starting up at the local Y, and she was contacting Wilson's past students to procure new members. The name of the club was Kamahoali, which the founding members claimed was Hawaiian for "great shark hunter." Link invited me to attend the next meeting. I did. I liked the people I met and I joined the club on the spot.

The club organized trips to local lakes, quarries, bays, jetties, rivers - in fact, to any place that had water deep enough for a person to become submerged. Once the treasury swelled with sufficient money from dues, the club chartered boats along the New Jersey shore in order to visit the old wooden scrap heaps and piles of metal rubbish known as shipwrecks. I really didn't care where we dived, as long as I could gain experience under water and become more proficient with my equipment. I went on every trip.

I was not overly impressed by the first wrecks I saw. They were so broken down after their years in the sea that great imagination was needed to discern form and structure in the thickly encrusted remains. They looked more like underwater junk yards than the hulls of stately vessels. In fact, a sunken tanker or freighter is no more than the marine equivalent of a smashed up eighteen wheeler after a horrible highway accident. Shipwrecks are bigger than truck wrecks, and usually result in a higher number of casualties, but other than the matter of degree, the only difference is that wrecks in the ocean are generally too big to tow to a yard for scrapping so they are left where they come to rest - out of sight beneath the waves and therefore out of mind.

An Unexpected Challenge

What I discovered immediately was the personal satisfaction I gained from meeting the challenges of diving. Each dive presented an unforeseen obstacle to overcome or offered some nuance previously unnoticed. Eager to experience the underwater world in its multitudinous forms, I joined a small group on a Caribbean venture to Andros Island, in the Bahamas. The significant feature of the trip (besides taking my first underwater photograph) was that, with only a couple dozen dives under my weightbelt, I descended "over the wall" to a coral ledge at a depth of 200 feet.

The excursion to depth was led by a guide who had several charges to watch. The plan was to descend as a group to a ledge at 185 feet, stay a couple of minutes, ascend slowly to 60 feet where the rest of the tourists were ogling fish on the reef, then swim slowly along the shoaling reef to where the boat lay anchored, at which point we would surface. No stops were planned for decompression because the leisurely ascent provided sufficient allowance for such a brief time at depth. I followed the guide to the appointed ledge, but when he turned toward the others, who required his attention, I dropped down deeper in a flush of exhilaration. I liked the feeling of depth.

Upon reading this one might conclude that the narcotic effect of nitrogen inspired my evident excitement, and at the time I might have agreed. But I've long since learned that such was not the case. The description of nitrogen narcosis as the "rapture of the deep" is the most exaggerated falsehood ever foisted upon the public. Throughout the years I've met numerous people who've described their sensations of narcosis. I have on occasion been "narked" myself. Not one single person has ever described narcosis as an overall feeling of euphoria. People become fuzzy, forgetful, lethargic, scared, disoriented, uncoordinated, but I've never met anyone who tried to give air from his regulator to a passing fish. Nitrogen narcosis is closer in equivalence to drunkenness than to euphoria.

I found in the depths a challenge that suited my nature: a place to go where others had not preceded me. Certainly my Bahamian guide had not preceded me, but he sure followed me in anger. I didn't need to see his jerking thumb to know that he wanted me to ascend - and *now!* I could see it in his face. I think he might have been better-natured about my slight disobedience had it not been for an incident the day before, when a panic-stricken teenage girl screamed straight for the surface from the same ledge and passed out on the way. She survived without hospitalization.

Three years passed before I dived so deep again. This was largely due to the lack of opportunity, as the only water of equivalent depth within sixty miles of the Jersey coast was the Mud Hole, where few charters ventured. But it was also due to my increased awareness of the potential danger inherent in such a dive equipped as I was without back-ups: only a single tank with one regulator.

I claim ignorance. If I made the same dive today I would be forced to claim stupidity.

Taking up Space

Early in 1972, new vistas of opportunity opened to me. Roach was then assuming full control of EDA. He expanded the dive schedule by taking more charters with the *Sea Lion*, whose proximity to Philadelphia was advantageous to him, and by phasing out the Long Island boats. The *Sea Lion* was on the way to becoming EDA's principal dive boat. Hoffman couldn't have been happier. But during this transitory phase Roach was having trouble filling the *Sea Lion* even though it was only a six-pack. Club membership was down, apathy among the active members was increasing, and the Long Island members didn't like the long drive. Since the charter fee was a set amount which was divided equally by the number of divers a boat was supposed to accommodate, any shortage or unpaid "spots" had to be paid out of the club treasury, which was fast dwindling. EDA needed new blood. I became part of the transfusion.

By that time I was willing to dive anywhere, on anything, with any group or club. Roach let Wilson know that he had spots available. Wilson told Link. She announced the news at the Kamahoali meeting. Three of us signed up and car-pooled to the boat. We were very definitely outsiders that day: no-accounts whose only value lay in making up the monetary deficit. The EDA members scorned us as novices.

On the other hand, Hoffman treated us as very valuable commodities: future customers. Often, as divers offloaded their gear at the end of the day, he declared his motto: "Come again. Bring money." I did.

I filled in on other days when Roach ran short. Those dives were invaluable learning experiences far superior to the dives I made with Kamahoali. The *Sea Lion's* cabin was the size of an elongated walk-in closet. The divers huddled shoulder to shoulder on benches behind Hoffman, who stood at the helm. Sharing this confined space with accomplished divers enabled me to listen to the conversation, not only about diving techniques but about the wrecks and their attractions. It also enabled me to observe skilled divers in action. I soaked up information like a sponge, learning by osmosis what was not taught in courses or written in books.

Roach and some of the EDA members were not particularly friendly. If anything, they were annoyed by the simplicity of my questions. But Hoffman and his mates were always helpful. Bob Archambault, a carpenter by trade, described the layout of the wrecks. Bill Hoodiman, an elevator constructor who worked with Hoffman as a helper, explained the intricacies of catching lobsters. John Pletnik, a man of the world, pointed out the advantages of double tanks, pony bottles, and other deep-diving esoterics. Danny Bressette, another elevator constructor who worked with Hoffman, provided gross entertainment with a constant barrage of offensive language that was intended to shock sensitive ears, by exposing his sexual anatomy, and by playing repulsive pranks such as chucking slimy sea anemones into his mouth. To this social atmosphere Hoffman added such distinctive phraseology as "like a turd from a tall cow's ass." Diving on the *Sea Lion* was never achromatic.

A NEW LURE

Once I found a brass porthole on the *Maurice Tracy*, a collier that sank in 70 feet of water after a collision in 1944. I tried to free it from the wreckage but had to give up when I ran out of air. Hardly had I mentioned my disappointment than Hoffman insisted that I take one of his tanks and go back down to work on it. He even loaned me a hammer and chisel. I still didn't get the cherished relic, but I knocked off enough encrustation to see that it wasn't a brass porthole at all but an iron rectangular window.

Here was another facet of wreck diving that appealed to me: I liked finding lost items. And soon I found other items that *were* brass, and brought them to the surface. Not that these items were worth anything, monetarily or otherwise. They certainly were not rare antiquities. Their sole value was intrinsic: they were simple souvenirs, mementos of successful dives, keepsakes whose usefulness had long since ended and under normal circumstances would have been trashed or melted down. These articles were worth something to me only because I found and recovered them myself. Left on the bottom, they were useless scrap.

I was an underwater trash picker, collecting worthless scraps of the recent past and returning them to society. These were items that I could have purchased in any nautical antique store. But a wreck-diver felt no more sense of achievement in buying maritime collectibles than an angler felt in buying fish at the market instead of catching his own.

It soon developed that I had an "eye" for spotting artifacts which countless other divers had passed over without noticing. My trifling childhood knack of finding coins on the ground became a priceless skill. Who would have thought?

Expanding Horizons

Meanwhile, whenever I tried to convince my newfound friends to help me explore the underground river in Blue Hole - my raison d'etre for diving - five minute's description of the sink hole and what it would take to lower scuba gear to the bottom of it was enough to disenchant the most intrepid of them. At the same time, the more wrecks I dived, the less I cared about cave exploring in general. Wreck-diving captivated me almost entirely.

In addition to the trips with EDA, I not only continued to dive with my newfound friends in Kamahoali, but I joined other Pennsylvania clubs: Aquarama Dive Club in Philadelphia, Main Line Divers in Ardmore, and CY Divers (Central YMCA) in Lansdown. When Roach at last opened EDA's membership to less qualified divers, I enrolled eagerly. Now I no longer had to stand in line for an available spot, I could sign up in advance for any trip I wanted. And I wanted them all. I dived most weekends during the summer season, sometimes both days with different clubs.

Disillusionment at Home

My wife made life at home intolerable. She was no Jane Wyatt playing opposite Robert Young, so my dream of a marriage like that depicted in *Father Knows Best* became a nightmare. I played my part by dutifully signing over my paycheck to my wife so that she could run the household and acquire the appliances, furnishings, and clothes she felt we needed. I had an allowance that I could spend on my activities. When money was tighter than usual, I worked extra hours. At one time I had three occupations going: my full time job with the union, driving a delivery truck in the evenings once or twice a week, and freelance employment on the weekends doing home renovations and installing 100-amp services. This was in addition to eight hours weekly electrical school plus homework. Still, my wife never had enough money. I couldn't figure out where it was all going. I just kept working and bringing home more.

In addition to this frenetic work schedule I found time to join the Mohawk Canoe Club, primarily to introduce my wife and son to the great outdoors. The club organized canoe trips and hikes to the local rivers and forests. Here began another side of my life subordinate to the thrust of this volume, for my associations in the club led me to other outdoor activities such as whitewater and wilderness canoeing, backpacking, winter camping, skiing, and mountain climbing. Perhaps of greater importance was my introduction to Jack and Rosemarie Schieber, lifelong friends and surrogate parents, and to many other good people whose love for the outdoors was superseded only by their love for their fellow man, a philosophy of life which I found attractive. Here was fellowship a-plenty to palliate the sting of the lack of affection at home.

I was never able to satisfy my wife's penchant for spending money, no matter how hard I worked or how many jobs I kept. I thought she wasn't doing her fair share to support our marriage. She developed such habits as sleeping until several hours after I left for work in the morning despite the baby's cries, leaving dishes piled so high in the sink that it was impossible to get a drink of water from the faucet, and piling dirty laundry on the floor all over the house despite the fact that we had our own washer and dryer. The house always looked as if a hurricane had just passed through it, leaving me to wonder what she did all day.

I found supreme happiness in raising a son. Michael was a never-ending joy and delight. I loved pushing him in the stroller along the sidewalks in the evening or, when he grew past that stage, taking him for walks and holding him by the hand as I ran errands for my wife to the neighborhood stores. Otherwise, marriage was a source of great mental distress, the most intolerable of which was not my wife's constant bickering but the lack of respect she displayed for my awful war experiences. She continually ridiculed me in public and in front of our friends by, for example, claiming that I finagled my way out of Vietnam by standing up during a firefight, waving to the enemy, and asking to be shot. Again and again I explained to her how much her cruel gibes anguished me, but she demonstrated neither remorse for her actions nor sensitivity about my past physical suffering and continued embarrassment.

My attitude toward marriage grew bitter. I endured my wife's abuse only for the sake of my son. Eventually, though, I proved too weak for the task. One night I came home from my second job at about ten o'clock, exhausted as usual. As was customary, I went straight from the construction site to my delivery job without stopping to eat. This usually meant that I got warmed up leftovers at home. But this particular night my wife did not even bother to take her eyes off the television screen when I straggled wearily into the living room. I asked what there was for dinner.

"Dinner was at six," she said. "If you wanted to eat you shoulda been here."

I heated up some food, ate in the kitchen by myself, waited until the program was over so I could have her full attention, then told her I wanted a divorce. She didn't believe me at first, but as the days passed and I retreated emotionally into an impervious shell, she realized my sincerity. There followed an acrid conflict that was painful for both of us. In the end she got everything she wanted but my continued support. I lost everything I had ever owned and worked for, even the recess money that my father accused me of saving. Yet it was worth the price for freedom and peace of mind.

I blame myself for not having had the strength to sacrifice my life for my family. I freely admit that I acted selfishly in wanting something out of life beyond servility as an economic provider.

A New Life Alone

While my ex-wife lived in a newly furnished house full of the latest gadgets and appliances, I lived in a cheap apartment without a stick of furniture. It took a year of such Spartan lifestyle for me to pay off all the credit card bills that she had run up. I slept on the floor in a sleeping bag. Cockroaches dashed over me during their nightly sojourns. I stored my clothes in cardboard boxes. For a sofa I unbolted the back seat of my car and placed it in the otherwise empty living room. My kitchen set was a discarded table and chair that I picked out of the trash. I ate and came to enjoy TV dinners, without, however, a TV to watch. I stowed my scuba gear in the bathtub, where it got rinsed whenever I showered, and kept my tanks in the front closet.

I didn't know that the heating pipes from the basement ran inside the back wall of that closet to the third floor apartment. I learned that fact when the high temperature in the confined space heated the tanks and pressurized the air they contained beyond the capacity of the burst disk. I never knew that people could evacuate a building so fast. There was a thundering stampede as occupants from the adjacent units tried to beat each other to the parking lot. One man shouted and banged on my door as he raced by. He didn't wait for a reply.

The cacophony was ear-splitting when I opened the closet door. A whoosh of dust engulfed me. The interior looked as if a cyclone had hit it: escaping air still swirled around the walls in counterclockwise fashion. Everything stored on the top shelf had been blown off and deposited on the floor in a heap, including some very heavy boxes of brass artifacts. There was nothing I could do but wait for the noise to die down. Then I casually walked outside to face the people shivering in the cold. They included one young mother cradling her infant in her arms. I explained the situation calmly and stressed the lack of danger. I repeated the explanation for the benefit of the fire fighters when the fire engines arrived.

Wreck-Diving Enthusiasts

With five club schedules to choose from I had opportunities to dive on every known wreck off the Jersey coast. From local lore I quickly learned which wrecks seemed the most interesting to explore, then signed up with whatever club offered trips with those destinations in mind. In this fashion not only did I learn about the wrecks, but I learned about wreck-divers - the good, the bad, and the ugly. My prolific dive club affiliations soon put me in touch with the foremost divers on the east coast, many of whom wore outlandishly large hats. Those with normal sized heads became friends. The others I tolerated.

The majority of divers were good, honest people and slice-of-life Americans representing a cross section of civilized humanity. They dived for fun, excitement, escape, or all of the above. General group dynamics were crystallized by camaraderie: most people enjoy sharing experiences with those who have similar interests. These people were not driven as much by wreck destinations as they were by club associations. They participated in club functions only, and would not dive a wreck they wanted to see unless that wreck was listed on their own club's schedule. They would not think of going there with another club.

The failure of the individuals of one club to associate with the individuals of another club fractionated the wreck diving "community" into discrete enclaves, and created separate ideologies almost maniacal in their fervor. This hatred of strangers, this lack of tolerance for alternative views, this refusal to acknowledge the rights of others, this us-and-them pathology, runs rampant throughout human history. It is the cause of war, persecution, and many other strifes which mankind has brought upon itself. Many early dive clubs failed to rise above the baser elements of cultural segregation. This led to strong rivalries and unhealthy competition between divers who should have been working together to promote underwater safety.

Dive Club Paranoia

The primary forces driving these rivalries were the most contemptible yet legally acceptable of human passions: the obsessive thirst for possession and control.

Possessionism in wreck diving regard can be characterized generally as the compulsive collection of artifacts. By artifact, as I have already explained but which I want to re-emphasize, I don't intend the meaning in the scientific sense - since items recovered from recent wrecks have no archaeological significance - but in the broader and more general usage as a manmade object which is constructed of material sufficiently resistant to the corrosive nature of the sea to survive for a while the processes of decay and deterioration. Technically, an iron object is an artifact, but wreck-divers don't consider it worth recovering, and therefore don't think of it as an artifact.

Aside from the ordinary desire to bring back a souvenir from a dive - whether it be a lobster or a brass junction box - the definition of wreck-diving possessionism can be extended to include historical information, physical description of a site, even the location of a shipwreck. A psychiatrically disturbed person desires not only to own and covet an object - from a specific artifact to an entire shipwreck - but wants to monopolize the object for his exclusive benefit, and in extreme cases he works actively toward excluding others from what he believes to be his own private domain.

More shameful than exclusive possessionism is the insatiable desire for domination. The majority of dive clubs were steeped in politics. All too often they were governed by control freaks who seldom had the interests of the club at heart, but who wanted to exercise power and authority over the membership. These people ruled through aggression and intimidation, they dictated policy, and they scheduled the destinations they wanted despite majority wishes to the contrary. The irony of the situation was that in many cases the club officers were neither the most skilled nor the most active divers. Their primary interest lay not in diving shipwrecks but in exercising authority. Each club was composed of rival cliques vying for command, and insiders who enjoyed special privileges from which outsiders were excluded.

In the early 1970's, a newly discovered shipwreck was practically a war zone. Long-standing dive clubs were enemy factions that not only competed against each other, but that fought over who got to a wreck first in the season to reap the new harvest of lobsters, or who got the most sought-after relics exposed by winter storms. A novice diver was quickly subverted by the disposition of the club he joined, and soon acquired the prejudices maintained by the club.

Kamahoali was an exception, undoubtedly because the club was new and its members were newly certified, and were not influenced by existing standards of behavior. The club's dominant feature was camaraderie.

Gaining Experience

For someone like me, who just wanted to dive, these were trying times. They were also fresh and exhilarating times.

I learned how to catch lobsters, how to work with tools under water, how to deploy a liftbag to send my artifacts to the surface. When a single tank no longer afforded me enough air for the length of time I wanted to spend on the bottom, I bought a cross-over bar and a backpack and converted my two singles into one set of doubles. When my friends did not share my enthusiasm for entering wrecks, I dived alone. In EDA, nearly everyone dived alone unless one had an objective that required assistance. Solo diving appealed to me because I could operate at my own pace.

I exceeded the no-decompression barrier with caution, not because I was afraid of decompression but because I didn't understand it. At that time and for the next couple of decades the certifying agencies assumed an ostrich posture toward decompression. Their attitude was: make believe it doesn't exist and maybe it'll go away. This unrealistic approach was as harmful to the diving public as ignoring the reality of venereal disease. The vacuum of ignorance left people totally unprepared to deal with the problem effectively.

Brush with the Bends

I got bent in 1972 after a fifteen minute dive to 140 feet, which according to Navy Tables required no decompression. I had just acquired a mechanical decompression meter, the latest innovation in decompression diving. The chief advantage of the meter over the Tables lay in its ability to account for time spent at depths shallower than the maximum depth attained. The Tables assumed that a diver remained at the deepest depth for the entire dive, and couldn't compensate or give credit for time spent on the upper part of a wreck with high relief. By means of a ceramic filter and an expandable bag, the meter calculated the *actual* absorption of nitrogen under pressure. This resulted in more bottom time or less decompression time, however you wanted to work it. The SOS meter was the first multi-level exposure meter.

Since I wore a depth gauge on my left forearm, I strapped the meter to my right forearm. The wetsuit material compressed at depth, causing the strap to loosen and let the meter slip out of place. I could have tightened the strap, but it was quicker to push the meter up my arm to where my arm was thicker. During my ascent, the wetsuit material expanded under the strap and restricted my circulation. On the boat my arm was bruised and sore, and the dull ache soon became torturous. I wondered what was wrong, but so little was understood about decompression injuries that no one, including a medical doctor on board, had any advice to offer.

I figured that I was suffering from some kind of decompression injury that was related to the tightness of the strap. I also figured that recompression was the best cure for the problem. When we moved to a second site at 120 feet, I made a conservative fifteen minute dive that required no decompression, and made a slow, drawn-out ascent but without staged stops. My strategy didn't work.

I scarcely got back on the boat when the pain became excruciating. With great effort and agony I stripped off my wetsuit jacket. My entire arm was swollen and colored dark purple, like a giant splotch from the shoulder to the fingers. I collapsed to the deck. Hoffman and his crew were working the wrecks for brass and bronze valves, pipes, and fittings - called "mungo" - that could be sold for scrap. He was busy directing recovery operations and pulling in lines while the mates were in the water chasing liftbags. I passed out from unbearable pain. When I regained consciousness, Hoffman told me to get out of the way. Heavy valves and lengths of pipe were being hoisted aboard and the center of the deck was needed for storage and working space. I rolled back against the tanks that were strapped to the gunwale. My mind was so dulled that I was barely able to comprehend what was going on around me.

As the activity died down I regained more awareness. Gene DellaBadia, a psychiatrist, examined my arm but admitted that my condition was beyond his medical experience. I wanted to go back in the water and see what it would do for relieving the pain and reducing the swelling. DellaBadia volunteered to go with me, to keep an eye on me in case I passed out. He helped be back into my wetsuit jacket by reaching up the sleeve from the wrist and pulling my paralyzed arm through.

We donned single tanks. Hoffman lowered a weighted line off the stern. DellaBadia and I rolled over the side and descended to twenty feet, where I got some relief from the pain. After an hour or so I felt

1970's: Decompression Comes of Age - 225

better. I could make a weak fist and bend my arm at the elbow. The pain decreased to a dull ache. We hung at ten feet till I ran out of air. My forearm was swollen so tight that DellaBadia couldn't make a dent in the skin with his finger, but the purplish coloration had disappeared.

It was dark by the time we reached the dock. Despite my incapacity, no one offered to help unload my gear. My right arm was nonfunctional, so I had to carry everything one-handed. I drove home with my arm limp at my side. During the night the pain returned with a vengeance. I woke up moaning. Toward dawn I realized that I needed medical treatment - I couldn't sleep it off - so I drove downtown to the University of Pennsylvania medical center, where a recompression chamber was available.

The emergency room clerk wouldn't let a doctor examine me until I signed an admission form. She didn't care that my right hand was swollen stiff and unusable. I signed in by using the fingers of my left hand to hold a pen between the fingers of my right, and by moving my right hand with my left. The result was hardly legible, but it suited her rigid mentality. I've often wondered how unconscious patients got past her booth for treatment.

I explained that I needed to see a decompression specialist. In short order Dr. Idicula arrived. He was interning in the States from India, but, as I came to find out, he had more to teach than to learn. He asked the minimum requisite questions while examining my arm, then without perceptible deliberation told a technician to prepare the chamber for operation. A power failure delayed proceedings for an hour. When preparations were complete, a technician escorted me into the chamber and sealed the hatch. Idicula peered in through a tiny porthole with a friendly smile on his face. He spoke through an intercom as the chamber was pressurized, giving instructions affably to me and the technician. I lay back on the bed as air whistled into the chamber.

The air was compressed to the pressure equivalent to a depth of 60 feet. Although the swelling was not substantially reduced, I felt immediate relief from pain. During the next two and a half hours I breathed oxygen from a mask, with short breaks on air. There weren't enough blankets in the chamber to keep me warm during "ascent." Expanding air reduced the temperature far below my comfort level. I felt much better when I emerged at the end of the "ride."

Idicula had his diagnosis ready for me. He told me that I did not have ordinary decompression sickness caused by nitrogen bubbling in the bloodstream, but lymphatic edema: swelling of the lymph system. Only two previous cases were known in the history of hyperbaric medicine. He showed me pictures in a book of a case nearly identical to mine. After assuring me that I would recover fully, he asked if he could take comparative photographs of my arms, that he could use to illustrate his report on the case. Always eager to help the advancement of science, I readily agreed.

I had always had trouble with my lymph system. Swollen glands or lymph nodes so alarmed doctors when I was eight or nine years old that they performed a biopsy on a gland in my neck. They thought the swelling was cancerous, but the biopsy proved negative. Throughout my teenage years and in the army I often had swollen glands. An antibiotic such as penicillin provided a quick cure.

Idicula recommended that I remain in the hospital for a few days for further observation, at least until the swelling was significantly reduced. He rigged up a traction device next to my bed. Whenever I was lying down my arm was suspended overhead in order to let gravity assist natural drainage. The doctor told me that when I was away from my bed I should hold my arm in the air whenever possible. Walking through the halls for daily x-rays I looked like some crazed neo-Nazi heiling the long dead Fuhrer. I had to maintain my sense of humor because so many people laughed when they saw my peculiar posture.

My arm was still slightly swollen at the end of the week, but Idicula released me with instructions to return the following week for a checkup. The accident and recovery interfered seriously with my social calendar, as I was slated to leave on Saturday for a two-week-long wilderness canoeing trip in Quebec. The infirmity of my arm rather than the doctor's appointment prevented me from departing as scheduled. I stayed home from work. When Idicula gave me a clean bill of health, I was raring to go canoeing. Fortunately, the trip was divided into two one-week segments on different rivers, so I was able to make the second week.

INTRODUCTION TO REMOTE DECOMPRESSION

One month after the accident I was back in the water. I had yet to make a bona fide decompression dive, but during the next couple of months I stretched my bottom time to the point where it was necessary to hang onto the anchor line for five or ten minutes before surfacing. It didn't seem like a big deal. But when I made a 155-foot dive which required thirty minutes of decompression - with no anchor line in sight - I was terrified. I understood the mechanics of decompression, I carried a home-made reel based upon the design in common usage, but I had never actually done a remote decompression.

The operation did not go smoothly. My heart thumped and my hands shook as I struggled to reach the reel on my back. The rope was unmanageable in fingers that were stiff with cold. I was too frantic to look for a suitable piece of wreckage with rounded edges. I had extreme difficulty controlling my buoyancy during the ascent. I couldn't maintain the proper stages because of current and the passing waves. My watch ticked away meaninglessly. What prevented me from surfacing was not a physical barrier but intellectual appreciation of the medical consequence of doing so.

After that awful experience I practiced decompression reel deployment even when its use wasn't necessary. I altered the method of securing the reel to my tanks so I could release it more easily. And I developed a solution to a flaw in the design of the reel. Sisal fibers swell when they absorb water. Because the line must be wrapped tight in order to keep it from coming loose on the spool, the combined swelling of adjacent strands forces the wraps at the ends to bulge around the dowels. These bulges snag when the line is unreeled. My contribution to the evolution of the decompression reel was the addition of a flat circular disc or plate at either end. This not only prevented the line from bulging, but it forced the strands to compress as the fibers swelled, thus compacting the line and making it less likely to come loose after its first soaking. The innovation caught on and eventually gained universal acceptance. It accounts for the present form of decompression reels that are manufactured today.

ADVANCED REMOTE DECOMPRESSION

Another phase in the evolution of remote decompression came about more slowly. In the commonly accepted practice of the times, the end of the line was tied to wreckage and the line was unreeled as the diver made his ascent. Buoyancy to hold up the line was provided by the diver: in the early days of wreck-diving, by the neoprene wetsuit material; later, by means of a buoyancy compensator or inflated drysuit. The diver then bobbed like a float. Often, after becoming stabilized, he was pushed down by itinerant currents, forcing him to let out more line to reduce the depth, at which point he was pushed down once more by the current, forcing him again to let out more line - and on, and on, and on. This yo-yo form of decompression was exhausting, it wasted air, and it was less than safe or efficient.

In the modern method, the end of the decompression line is not tied to wreckage. Instead, a liftbag secured to the end of the line is inflated and sent to the surface while the diver holds onto the reel and lets the line unspool. When the liftbag hits the surface, the line is cut from the reel and tied to wreckage, creating in effect a personal anchor line.

The first person I saw do something similar to this was Tom McIlwee, a printer. He wrapped thin nylon line on a miniature dowel, and used a sausage-sized marker buoy inflated from a CO_2 cartridge, all of which was compact enough to carry in a small goodie bag. He popped the cartridge on the bottom, let the inflated marker buoy pull the line to the surface, cut the line off the reel and tied it to wreckage, then ascended.

I rejected this method at the time, and resisted employing it when it came into vogue in its later incarnation. My reasoning was that the longer stay on the bottom - while waiting for the inflated marker to reach the surface - increased decompression time, or, if anticipated, decreased bottom time. Also, the thin line could chafe apart easily. And furthermore, the small marker that McIlwee used - all that was available then - would not inflate below about 100 feet due to the small size of the CO_2 cartridge, and did not have the capacity to hold up the line in a strong current. The mechanical problems were overcome by replacing the marker buoy with a liftbag, as noted above, and by employing a sisal-style decompression reel.

Toward the end of the decade, Jon Hulburt and I sent up so many artifacts on liftbags that we got into the habit of ascending the safety line (tied to wreckage so the artifacts would not drift away). This enabled us to cut the liftbag free upon completion of decompression in order to swim the artifact to the boat. In the process we discovered how comfortable the decompression was: the stability of the line and the ease in maintaining the proper stages for decompression far outweighed the disadvantage of the few minutes added to the ascent time. (Hulburt was a chemical engineer and one of wreck-diving's greatest innovators.) During the winter, we discussed using the same procedure even if we had no artifacts to float.

I was still slow to adopt the method. For a while I stubbornly refused to send up a liftbag for decompression purposes unless the action was necessitated by the recovery of an artifact. Eventually, though, after a few difficult hangs in strong current and rough seas, I gave in and adopted the procedure full time. This is a good example of how a mindset can make one resistant to change.

It is informative to note that McIlwee first demonstrated his concept to me in the early 1970's. Not until the late 1970's did the method gain large-scale acceptance, independent of McIlwee's prior application. Once a sufficient number of divers adopted the idea, the old technique was phased out and forgotten. Upcoming divers knew nothing else. The development thus was complete - until someone invents a more suitable mechanism.

Ironically, while going through de Camp's unpublished photos from the 1960's, I noticed a picture of a diver carrying a sisal-style reel with a vest-type BC attached to the line. The idea was similar to those described above, although de Camp assured me that the system had never actually been deployed. This all goes to show that practical discoveries and improvements may occur and recur until the appropriate time arrives for their acceptance.

"If at First You Don't Succeed . . ."

There were failed ideas, as well. In 1972, John Pletnik was just about the only non-commercial diver in America who owned a drysuit. He purchased it in Sweden during his travels abroad. Drysuit diving was a concept I was willing to embrace without exhortation. In winter and spring, the water temperature barely rose above the freezing point. Dive boats were unheated. Sometimes we had to shovel snow off the deck before loading our gear. On the ocean, spray turned to ice which caked all exposed surfaces. We had to chop the ice off the manifolds before the regulators could be put on the tank valves. We didn't dare test our regulators before entering the water because the condensation from moisture in the exhaled air froze and caused the second-stage diaphragm to jam open, which made the regulator freeflow. Wetsuits offered scant protection from the elements. I shook violently during the dive, breathed air at an enormous rate, and couldn't get warm during the surface interval. Those, as they are wont to say, were the good old days.

As soon as drysuits were available on the open market, I bought one. Mike de Camp was slow on the uptake, but when he saw me luxuriating in relative comfort, while he shivered in his 3/8-inch neoprene wetsuit, he decided to give it a try. He had so little faith in the watertight integrity of the seals and zipper that he wore his wetsuit underneath instead of longjohns, figuring that if the drysuit leaked he had an adequate back-up. He bobbed on the surface like a cork. Even wearing two weightbelts he couldn't get down. Not only did he have to abort the dive, he had to be towed to the ladder because, like a child put out to play in the snow, he couldn't move his arms or legs. We all had a good laugh over his discomfiture. Well, all but one!

I took a different tack. Reveling in the inflatable buoyancy of the drysuit, I thought of a way to use it as a secondary liftbag for bringing small objects up from the bottom. Around my neck I wore a loop of nylon rope to which I secured a leader with a large aluminum carabiner. When I found a porthole on the *Arundo* I clipped it to the carabiner. I couldn't find the anchor line in the dim visibility, so I tied off my decompression line and inflated my suit to carry me and my prize to the surface, 130 feet above.

Tom Roach and Jan Nagrowski happened by just at that moment. They couldn't find the anchor line either, and each of them was carrying a porthole. They liked my neck loop idea so well that they decided to try it, too. The problem was that they didn't have neck loops. They used mine instead. I was already ascending with both hands holding onto the ends of my reel. I shook my head when I realized what they intended, but they ignored my protestations and clipped both portholes to my neck loop. The weight dragged me down and created slack in the decompression line. With extra effort - and still holding onto the reel - I pressed the chest inflator button with my thumb. I regained positive buoyancy, got the line taut, then let more sisal off the reel. Halfway to the surface my drysuit ballooned with expanding air that was offsetting not only my weight but the weight of the three portholes, a situation which exceeded the design limitations of the system.

The drysuit arms blew up so big and hard that I couldn't bend my elbows to reach the deflator button with my thumb. Nor could I wrap my legs around the decompression line because the drysuit legs were so distended. The final disaster struck when the weight of the portholes pulled down my head and caused a slight shift in balance. Air rushed into the legs of the drysuit and flipped me like an egg.

My situation was desperate. I hung completely upside down, gripping the ends of the reel with all my strength, and about to take off for the surface like a missile launched from a nuclear submarine should I lose my grip on the reel. I was helpless. Fortunately, Roach and Nagrowski observed my plight and realized that if I let go, the reel and line would fall back to the bottom, and their own decompression would be compromised. It took both of them hauling on my feet to invert me. After such dire straits I rejected the neck loop as an unhealthy invention.

A Lesson in Wreck Penetration

Roach's hard work to expand EDA's schedule and increase membership gained great momentum in 1972. In an effort to attract more dues-paying members to the club, he reduced the number of deep dives on the schedule and concentrated instead on wrecks within ordinary reach, peppered with a few destinations more provocative, such as the *Texas Tower*, the *Bass*, and the *U-853*. Hoffman took advantage of EDA's growth by upgrading to a larger boat, also called the *Sea Lion*. He could now carry twelve passengers uncomfortably.

The *Texas Tower* was a U.S. Air Force radar installation which collapsed in a storm in 1961 some sixty miles offshore. The depth to the bottom was 185 feet, but since it rose to within 65 feet of the surface it made a good dive for all skill levels. The story of the *Bass* has already been recounted. The *U-853* was a World War Two German U-boat that was depth-charged by U.S. Navy destroyers in the final days before Germany's capitulation. Its depth was 130 feet.

Roach and I dived together inside the *U-853*, whose interior at that time was largely unexplored. I stood safety for Roach as he tried to squeeze through the conning tower hatch. He couldn't fit through the opening despite having removed his pony bottle. Instead, we entered the wreck through a large damage hole in the stern, which gave access forward to a hatchway into the motor room. This hatchway was a watertight hatch and a pretty tight fit. Roach barely squirmed through and I didn't find the going much easier. My tanks banged against the upper lip and my lead weights scraped both sides. After the restriction we swam through the compartment and through another hatchway into the engine room. The water was clear for Roach, who led the way, but visibility was limited for me because I followed in his fin wash. Roach stuck his head through the control room hatch, then signaled to turn around.

On the way out I was in front. Rusty particulate matter fell from the overhead like dense red snow. The space between the engines was narrow. When I reached the aft hatchway I went straight through into the jumbled mass of wreckage in the room adjacent to the outside exit. Roach wasn't with me. I turned and shone my light into the gloom. When he didn't appear after ten or fifteen seconds, I felt my way through the murk to search for him, thinking that he must have gotten stuck in the hatch.

There was no sign of him. The compartment beyond was a black void occupied only by swirling mud and silt, like an inkwell filled with sawdust. Just as I poked my head into the opening I caught a glimpse of fin tips overhead. I pulled on one, and down came a pair of legs. When Roach's face descended from above, it showed stark white against the pall from behind. He had had a few tense moments.

Later I learned that he lost sight of my light sometime during our return, after which he bumped into a bulkhead that hadn't been there before. A submarine is a hollow tube crammed with machinery and having no side passageways. Getting lost is not as much of a concern as getting entangled. He was disoriented by the presence of a bulkhead where none was supposed be, and which he soon discovered was bounded on either side by perpendicular bulkheads. Visibility that was already poor dropped to nil.

In actuality, during his return through the motor room he became a bit too buoyant. He arrived at the doorway a couple of feet above the lintel. Instead of seeing the opening, he saw the bulkhead above it, and never thought to look below.

Getting Ahead – Too Far Ahead

Despite these and other adventures that I shared with Tom Roach, and all the time we spent together driving to and from the shore, our relationship never kindled to the warmth of true friendship. This was because he believed in strict hierarchy: an inviolate pecking order based upon time in grade. To his way of thinking, a person's rank was established at the moment of certification, from which point he advanced according to his position on the scuba conveyor belt. Roach looked up to EDA's old guard, treated his contemporaries as peers, and dismissed me and other neophytes as underlings who were permanently marooned in a subordinate echelon with no chance of leapfrogging ahead of our station. This attitude was pervasive throughout the club. To me, hierarchies are artificial constructs completely without validity. They are designed and embraced by people who are intimidated by equality.

I can point with precision to the very second when Roach first felt threatened by my improvement in ability and potential advancement: it was when I climbed aboard with a porthole from the *Tolten*, one that so many people had tried to get off and had given up as impossible. In truth, my success where my "betters" failed involved no great amount of insight. Those before me, being right handed, pounded on the hinge pin from the side that was natural to them, without taking the trouble to scrape off the encrustation to examine the ends of the hinge. Once I did so and determined that the pin was capped and needed to be driven out from the left, I had the porthole free in a trice.

I displayed my prize proudly, expecting praise. Instead I received a lackluster look and a forced, "Oh, you got it." I made matters worse soon afterward when I recovered the "impossible" window from the *Stolt Dagali* by hacksawing through two hinge pins that had proven unmoveable by means of a sledge hammer and drift pin punch because both ends were peened over.

Roach needed to be the kingpin. My constant upstaging rankled him to no end. He expected me to stay in my place, not to leapfrog over my predecessors. It might be said that we had the classic love/hate relationship, except that such a perception was purely one-sided. I was just doing my own thing and doing it to the best of my ability. That the successful endeavors of a novice upstart stung him where he was hubristically sensitive was not my intent. He sowed the seeds of his own discord.

Back Seat Supervisor

In this light, 1973 was a year of growth, extremes, and contradictions. Despite Roach's increasing presumption and tendency toward faction, the large core membership of EDA was molding itself into a happy holistic fellowship that was larger than the sum of its parts. I became the club's titular second in command and the principal liaison between the serfs and the "king." Roach's leadership by domination coupled with his lack of imagination limited the club's development. I saw EDA's true potential being wasted in his hands. Yet I never tried in the least to wrest control of the club. I was content to make suggestions, he was willing to implement them.

Thus when he complained about members who defaulted on payment for missed dives, I suggested instituting an escrow fund.

According to my system, all members (myself included) deposited in the club treasury an amount of money equal to the cost of a dive (which at that time, believe it or not, was $15). If a person failed to show up for a dive and send payment in kind soon afterward, the spot was paid for from the escrow account, and the person was automatically taken off future charters until he made up the deficit. This way the deadbeats were eliminated. In theory, all escrow moneys were returned at the end of the year, but in practice most people "let it ride."

The Wreck-Diver's Code

A philosophy I campaigned for heavily was a code of ethics among divers concerning the recovery of artifacts. In contravention to the law of G. Magnus, I proposed a protection system in which an artifact's initial discoverer was granted the right to "work" the artifact without interference. The rules were fairly simple and were based upon elementary humanitarian precepts. In its purest form, an artifact's discoverer announced his discovery and described its location in order to differentiate it from similar artifacts in the vicinity. To prevent misunderstanding, the artifact could be marked such as with a line or a liftbag, or by leaving some tools on the site. The discoverer could then return to the artifact on succeeding dives without fear of someone jumping his claim in the meantime.

The only proviso in the code was the discoverer's responsibility to work the claim continuously in order to maintain his rights. The discontinuation of work constituted abandonment, in which case the rights were forfeited. This stipulation prohibited a diver from marking a multitude of hard-to-get artifacts just to keep others off them while he went exploring for easier ones. You either worked the artifact or you let someone else work it.

Since this unwritten code was based on the honor system it was (and still is) subject to abuse. There are those who want to have their rights respected without having to respect the rights of others, who agree to a code of ethics only when it is to their benefit to do so. A code of ethics which is based upon the convenience of the moment is no code of ethics at all, but a mockery of morality. By definition, a code of ethics is inflexible.

While most everyone agreed that the law of G. Magnus was a base holdover from primitive mentalities, and that civilized principals of conduct should prevail, there were those who used the system only to their own advantage, or to their own club's or clique's or group's advantage, and did not extend righteousness beyond their own arbitrarily defined sphere.

Testing the Code

The focal point of one controversy was Bill Nagle, an auto mechanic who had recently joined EDA but who had been diving for several years with Main Line Divers. Nagle conceived a bold plan to recover the main magnetic compass from the conning tower of the *Bass*.

The compass was secured to gimbals inside a bronze watertight housing which, even after all these years underwater, remained unflooded. The compass card was visible in its bubble of air through the glass viewing port of the binnacle cover. Nagle unscrewed some of the retaining nuts, of which there were about twenty, in order lift the cover from the binnacle so as to reach the compass. This took quite a bit of work because the nuts were heavily encrusted. He worked on the job for two dives straight but still had several nuts to remove when he ran out of time.

Not one to give up on a project half done, on the way back to the dock he talked with Brad Glas, captain of the *Helen II*, about chartering the boat so he could complete the job. Roach was incensed at Nagle's effrontery: first for having tackled a job that no one else had thought possible, and second for trying to "charter the boat out from under me." By this latter statement Roach expostulated his belief, if you can follow the reasoning, that since he made the initial charter arrangement with Glass, no one else had the right to charter the *Helen II* because the boat was his exclusive domain.

In any event, Nagle signed up for the next EDA trip to the *Bass*, whereupon he picked up the project where he left off. During his first dive of the day he removed the remaining nuts, but when he tried to pry off the binnacle cover with a crowbar, the cover moved only an inch or two then refused to budge any more. He ascertained the problem before he ran out of bottom time: the thick electrical cable that ran from the light in the cover to the chaseway in the binnacle did not have enough slack to let the cover pivot back on his hinge. Either the cable had to be cut or the hinge pin had to be driven out.

During surface interval I overheard Roach talking with several of EDA's old guard, including Don Nitsch. They agreed to jump in the water before Nagle and wrest the prize from his grasp. I didn't think that was fair. In addition to originating an effective plan for removing the compass, Nagle had expended considerable effort in carrying out the task. The others did nothing but watch like vultures until the feat was all but accomplished - and that after they scoffed at Nagle claiming that his proposal was preposterous and couldn't be done.

I pulled Nagle to a deserted corner of the boat and told him what nefarious plan was afoot. He couldn't make his second dive yet because he had been last in and out of the water. He needed more surface interval. I volunteered to complete the job and give the compass to him. If he harbored any feelings of mistrust he kept them to himself. He gave me the tools that I needed for the job and told me in detail what had to be done. The latter was important because at the time I had no idea how a compass was secured to its gimbals, or even what gimbals were.

Frank Messina, an appliance repair person, agreed to dive with me and help by holding a light. We entered the water with no one else but Nagle wise about our mission. Messina went into the conning tower from aft, I squeezed in through the narrow opening forward. We met at the compass binnacle in the middle. As I crouched on my knees, I pounded on the end of the pin with a hammer till it was flush with the socket, then used a pin punch to drive the pin through the hinge and out the other socket. The job was unexpectedly easy. I applied the crowbar to the side of the cover with the hinge, pried up the cover enough to get my hands underneath, and, with my back braced against the bulkhead, I heaved the heavy bronze cover off the top of the binnacle.

I expected the cover to fall with a clunk to the deck. Instead, it swung on the cable like a wrecking ball, slammed into my chest, knocked the air out of me, and pinned me against the bulkhead with its 150-pound mass. I struggled to extricate myself while Messina watched helplessly from the other side of the binnacle, illuminating my difficulty. Quite a few seconds passed before I was able to work my hands in a position of leverage. I shoved the cover inch by inch across my chest until I managed to slip out from under it. Then I gasped for air like a fish out of water.

Once I had my breath back I used a fine screwdriver to remove the retaining screws from the two gimbal rings. Again the job was easy. It was just time consuming. The most difficult part was fitting the tip of the screwdriver into the narrow slot of the screw. Messina was hard pressed to hold the light in the most favorable position. The screws and the gimbal rings were brass, so no corrosion bound the two together. I scooped out the compass and plunked it into the mesh bag that Messina held open. We attached a liftbag and sent the works to the surface. Nagle was fully suited. As soon as the liftbag hit the surface he jumped overboard and swam for it. He wanted no help from anyone who might lay a claim of ownership.

On Nagle's next dive he recovered the two gimbal rings and associated screws, and cut through the electrical cable with a hacksaw. He humped the heavy bronze cover to the forward opening of the conning tower, but was forced to leave it there when he ran short of time. He recovered the binnacle cover on his own charter a month later.

Roach was apoplectic. He didn't like those he considered inferior taking the glory for the day. The half dozen portholes I recovered over the next two months did nothing to mollify him about my encroachment. He and Nitsch got even with me, ironically, on the *Bass*. There was only one porthole left in the conning tower, and I started working on it. On another trip to the wreck, Roach and Nitsch went in to set the hook. I reminded them that I was working on the last porthole. They said to my face that they would leave it alone, but they were lying and knew that they were lying. They intended to recover the porthole just to spite me. They did, but they created great dissension in the ranks, and the members began to take notice of the darker side of their leader and his cronies.

Helm Recovery, or Ignoring the Hierarchy

But the point at which Roach recognized me as a veritable threat to his hierarchical belief system was when I recovered the helm stand from the *Ayuruoca*. It sat atop the wheelhouse wreckage in plain sight, and many a diver put his hands on it to try to shake it loose from its mount. No one told me it was impossible to remove, so I took down some wrenches and unbolted the six fat nuts that secured the base of the pedestal to the deck. I never said a word about tackling the job, so the first that anyone knew of it was when the liftbag hit the surface with the 200-pound helm stand suspended below it. Surprise!

This successful solo recovery was directly responsible for the unanticipated help I received on my next project. Two large steel freighters lay in close proximity to each other in the Mud Hole, at a depth of 200 feet. Historical sources indicated that they were the *Choapa* and the *Ioannis P. Goulandris*, both sunk in collisions during World War Two. But which was which? They were referred to collectively as the "Junior," without distinguishing one from the other. Roach obtained photographs of both vessels, but the visibility was so bad that descriptive features were hard to piece together. I proposed to recover the helm and its stand from the bridge in hopes of finding the manufacturer's name stamped in bronze. Since the *Choapa* was built in Denmark and the *Ioannis P. Goulandris* in England, a distinguishable name could very well establish the wreck's identity. Furthermore, the artifact itself was more desirable than the one from the *Ayuruoca* because this one still sported an intact wooden wheel complete with spokes.

When I told Roach what I intended to do, he offered to lend his assistance, figuring, I suppose, that second fiddle was better than no fiddle at all. I no longer needed his approval, but I was glad to have the help of one with his degree of expertise. Together with Danny Bressette we scouted the wreck one cold, December day. We spent the first dive swimming in the wrong direction without finding the bridge. This was partly due to poor visibility - it was virtually a night dive since no ambient light reached the bottom - but largely to our unfamiliarity with the wreck.

Maximum depth attained was 175 feet, total bottom time was seventeen minutes, after which we spent twenty minutes decompressing. This was a typical deep-dive profile for those days. On the second dive we located the bridge. I took compass bearings, examined the six nuts at the base of the pedestal, and recovered a loose porthole - all in twelve minutes. The most important knowledge gained was that the wreck lay canted to port with the starboard rail high. This aided greatly in future orientation.

After the dive I felt kind of funny. The fingers on my right hand were numb, and my right leg felt as if it were asleep. I mentioned these strange sensations aloud, but no one had much to say about it. By continually flexing my fingers and making a fist, and by walking back and forth on the crowded deck, I worked out the "kinks" until my feeling returned to normal. I soon forgot the incident.

Two weeks later found us back on site. Roach and I went down a shot weight that slid past the rail on the high side of the wreck. It took but a moment to get oriented. We didn't know exactly where we were, but we knew in which direction to swim. A minute or two later, after swimming some seventy-five feet, we gained the bridge, mounted the superstructure, and settled down next to the helm at 170 feet.

We worked together with peak efficiency. I scraped the encrustation off the nuts and loosened them with a pipe wrench, then Roach spun them off the rest of the way by hand. In short order we removed the nuts so that the threaded bolts were exposed. But when we pulled up on the stand to lift it clear of the bolts, it wouldn't come free. We pushed, pried, and beat on the pedestal, all to no avail. Up to this point everything had gone like clockwork. Now we were stymied by a pedestal that refused to budge.

A close examination revealed that the pedestal base consisted of a two-part assembly. A bronze connecting shaft that was two inches in diameter protruded from the sidewall, and at one time extended under the wooden deck which was now eaten away. We later learned that the assembly contained meshed gear heads which transferred the movement of the helm to the shaft and, ultimately, to the steering engine. The base plate and the shaft held the pedestal securely in place.

At the end of seventeen minutes I tied my decompression line to the wheel, and we ascended slowly. We hung for twenty minutes, then swam the line to the boat and handed up the reel with instructions to tie it off so we could use it to return directly to the job site.

Archambault and Hoodiman, mates for the day, decided to lend a hand without telling us. They descended our line to the bridge. While Hoodiman went to work on the shaft with a hacksaw, Archambault scraped the encrustation off the base plate and discovered four additional studs that Roach and I had missed. He then ducked through a doorway under the wheelhouse deck into the room below the stand. Here he found the nuts that secured the pedestal's lower flange to the deck. He removed all four nuts, which were overhead: a difficult job at best. By that time Hoodiman had cut through the shaft. Together they lifted the stand off the studs; it fell over onto the sloping deck. Quickly they secured a liftbag to the stand. They didn't have enough air or time remaining to fully inflate the liftbag, so they wisely left the stand where it lay and reported its condition.

The plan for that day's trip was to make one dive on the "Junior" and the second dive on the *Pinta*, which lay in shallower water, on the way home. With success imminent, Roach and I were not to be deterred from completing our mission. We made a short dive to complete the task. We dropped down the decompression line to the helm. The partially inflated liftbag prevented the stand from sliding down the slanted deck and over the edge to the muddy bottom thirty feet below. While Roach secured a second liftbag, I cut away entangling monofilament and attached another decompression line to the wheel. Roach purged air from his regulator into the mouth of the liftbag. We backed out of the way as the helm lifted with agonizing slowness off the deck, stood vertical, then headed sluggishly for the surface.

When my line stopped unreeling I cut it and tied it to the rail. Now there were two safety lines on the helm: one to the wreck and one to the boat. If the stand sank we could relocate it easily. We climbed up the decompression line. Our bottom time was ten minutes. We did no decompression.

After ascertaining that the boat line was secure, I severed the bottom line. It took little effort to haul the prize to the boat, but it took incredible exertion from everyone on board to lift the heavy stand and wheel out of the water and over the gunwale onto the deck without scratching the bronze surface or damaging the teakwood wheel. We wasted no time cleaning the encrustation off the rudder indicator on top of the pedestal head, and were overjoyed to see that the manufacturer's name was deeply etched in bronze: Robert Roger & Co., Stockton-on-Tees. The *Ioannis P. Goulandris* was built in Stockton as the *Eggesford*. The wreck's identity was firmly established.

Lost Inside

Enough air remained in my second set of doubles to make a 90-foot dive, so when we anchored on the *Pinta* I went down again, this time alone. I dropped through a doorway in the stern of the wreck into a narrow corridor barely high enough to squeeze along, scraping all the way. After fifteen or twenty feet I came to a transverse passage which I descended to the bottom. Then I squirmed forward a couple of body-lengths along a now-horizontal shaft and emerged in the low side of a small room full of mud and loose debris. I had been there before. I wanted to explore the area more thoroughly.

I moved carefully through the darkness, able to see only the spot that was illuminated by the beam of artificial light. The room was actually a centerline passageway from which doorways led to rooms on either side. Because the wreck lay on its beam, half the rooms lay below and were filled to the rim with mud. The rest of the rooms lay above. I swam forward slowly, poking my light and my head through the doorways overhead, and scrutinizing the cubbyholes. After twenty feet or so my apprehension grew stronger than my desire to continue.

I worked my way back till I reached the after bulkhead, through which a small square hatch led to the after steerage compartment. I shone my light through the opening, but I didn't want to go in there. The long corridor beyond was black and uninviting.

It was time to leave. I dropped to the bottom by the corner near the after hatch where the horizontal shaft led out. Despite my care in moving - I barely fluttered my fins - suspended silt obscured the slender opening. I knew the precise distance and direction from the after bulkhead junction, yet each time I felt with my hands for the hole, I bumped into solid metal. Again and again I tried to find my way out. In a short time I was panic-stricken.

Wild thoughts rushed through my mind. After recovering such a notable artifact from such a depth, I would not get to bathe in the glory of the achievement. Such a notion has no survival value and in fact is counterproductive. Yet I must admit that the thought did arise. I was aching to know how much air I had left - how long I had to live - but I resisted looking at my pressure gauge. Instead I shifted into full survival mode, intellectualizing in an instant that the knowledge could not help my plight in the least and that the time spent fumbling with my light and the gauge was better spent searching for the way out. I consciously controlled my breathing rate. My mesh bag kept hanging up on debris, causing me to lose valuable seconds. I wanted to pull the lanyard off my wrist and throw away the bag. I was about to do so when I tried a different tack first.

The water was clear at the top of the room but black as pitch at the bottom. I rose above the swirling silt and relocated the bulkhead junction that was my orientation point, then flipped my body around as if I had just entered the room, and felt for the opening with my feet. In this manner I was better able to conceptualize the proper position with respect to the bulkhead junction. I dropped to the oozy bottom and *backed* out of the room. I didn't know my plan was working till my elbows hit the sides of the shaft.

With my tanks and weightbelt scraping metal, I squirmed backward blindly until I could go no farther. Then I rolled over onto my side and looked up. There was the vertical corridor leading to the high side of the wreck. I climbed up quickly, traversed the outer passageway with incredible celerity, and slipped out of the doorway into open water. My fear evaporated at once. Only then did I look at my pressure gauge: my tanks had 500 psi remaining. Feeling at ease, I spent another five minutes cruising the wreck and looking for lobsters.

Preservation and Curation

Because four of us worked together to recover the helm and stand, I suggested that we share its ownership. This was found agreeable. I took the artifact home. In my basement I removed the wooden wheel from the gear casing in the head of the bronze stand so the different materials could be preserved separately. I took on the relatively easy task of cleaning and polishing the stand by soaking it in a barrel of muriatic acid, then leaching out the chlorides with a succession of fresh water baths, and polishing the outer surface to a golden tint with a fine wire brush mounted on a 1/4-inch drill.

Archambault undertook the prodigious job of disassembling and preserving the wheel. First he unscrewed the two circular brass retaining rings, one from each side. Next he separated all the curved connecting pieces from the outer band. Each piece was split lengthwise with one rough inner facing and one smooth outer facing. Then he removed each spoke from the bronze hub. He labeled each piece of wood, and treated each piece by rubbing linseed oil into all the surfaces. When the process was complete, many months later, he reassembled the wooden parts on the hub and transferred the whole wheel to me so I could install it on the stand.

This lengthy process of preservation was typical of the care and treatment that wreck-divers bestowed upon their artifacts.

The helm's early life was peripatetic. I wanted it put on public display where people could see and appreciate it. While I searched for a permanent location, the helm was displayed at two large shipwreck exhibitions, one at the Ferry Boat in Brielle, New Jersey, another at the Philadelphia Civic Center. Finally I negotiated an arrangement with the curator of the Philadelphia Maritime Museum to display the helm in a prominent place in one of its exhibit halls. The curator left the museum's employ shortly after I signed the papers that defined the terms of our agreement.

The curator's replacement was not as enthusiastic about the helm as his predecessor. He hid it out of sight in the museum's basement. I kept looking for the helm every time I visited the museum to do research, but never saw it. After a year, I asked the new curator for a status report. He told me that the helm was being kept for "study" purposes. That the museum wanted to possess the helm but did not want to let anyone see it contradicted the reason for placing it there. I calculated that more people could see it in *my* basement than would see it in the museum's basement, where only the staff were permitted to go. I pointed out the clause in the agreement - a clause that I had insisted upon - which stated that the helm was on loan for the purpose of being exhibited, and said that if he did not display it, I would be forced to remove it. He didn't want to display it, so he grudgingly "deaccessioned" it. "Deaccession" is a pretentious museum term that means "give it back."

Archambault came to the rescue. After I reclaimed the helm from curatorial oblivion, he found a home for it at the Admiral Farragut school, which displayed it prominently in the main lobby. It has been on public display ever since.

Body Recovery

Two other events from 1973 are worth recording, as well as one non-event. The events first.

On an Aquarama trip scheduled to go to the *Cherokee*, a gunboat that foundered in 90 feet of water in 1918, I learned at the dock that a diver had died on the wreck the previous weekend. His name was Gary Ford, and he was a member of a Delaware dive club. Tom McIlwee and a friend showed up that morning and invited themselves along to search for the body. Don Cramer, captain of the *Miss Shelterhaven*, asked those of us from Aquarama not to object. We did not.

McIlwee was a body collector. He tallied corpses like notches on a gunman's pearl handle. Whenever someone drowned in a river, lake, or quarry, McIlwee volunteered to recover the body. There weren't many diving fatalities, despite the media's persistent effort to make the activity seem deadly and death-defying, but whenever someone died in the water, McIlwee went out of his way to bring back the dead, strangers though they be.

I didn't think there was much chance of finding the body on the wreck. The ocean currents had had a week in which to waft it away. McIlwee thought otherwise. According to Ford's buddy, when last seen taking his final breath, Ford was struggling to remove the lanyard that secured his mesh bag to his wrist as the heavy brass shell casing in the bag dragged him down. McIlwee theorized that the weight took the body straight down into the hull, which was intact except for the broken bow and which was shaped like an old-fashioned bath tub from which the corpse could not escape. He cautioned us to go about our own business and let him handle the body search and recovery.

Visibility that day was about ten feet. My buddy was Frank Messina. We went directly to the bow where I had previously found a capstan, and spent the dive freeing it from the surrounding wreckage and rigging it for recovery. McIlwee and friend crisscrossed the hull but failed to locate the body. Joe Bianculli, to his great chagrin, stumbled over the dead diver by accident, then screamed to the surface in a fright. He was visibly shaken when he described its whereabouts to McIlwee - so shaken, in fact, that he got right confused with left.

Accidentally misled, McIlwee went in the wrong direction on the second dive. Messina and I, returning to our work site by a different route outside the hull, and thinking that the body lay the other way, were taken by surprise when we came across the corpse jammed face down on the sand under the upward curving steel plates where the prevailing current held it. I didn't set out to look for the cadaver, but once found I accepted the responsibility of recovering it.

I pulled the body out into the open and rolled it over onto its back and single tank. The mask was flooded with water and blood that hid the eyes from sight. Fish or crabs had eaten away the upper lip, exposing the front teeth in a gray grisly grin. The wetsuit had protected the body from further digestive abuse. There was no sign of the mesh bag or the death-dealing gun shell.

I had seen far worse in Vietnam, but Messina's background was more innocent. Not that he shirked from the unwanted duty, but he maintained a respectable distance while I performed the gruesome task of trussing up the body for the lift to the surface. It would have been an easy chore to tie a rope around the tank valve, but I was afraid that if the body slipped out of the harness it would be lost. After all, the object of recovery was the body, not the tank. I had to press myself against the corpse's chest in order to weave the rope under the backpack. I cinched the rope tight under the armpits and tied the ends in a knot in front.

Messina was justifiably horrified, but not so much that he couldn't help. Because I was saving my liftbag for the capstan, I signaled for him to give me his. I secured his liftbag to the knotted rope and purged enough air into the opening to take up the tension. Then I signaled for Messina to give me his decompression reel. The body was more important than any artifact, so I wanted to ensure that if the liftbag broached on the surface and the body sank, it could be relocated. I tied off the decompression line, handed the reel to Messina, and added air to the liftbag till the body lifted off the bottom. When the line stopped unreeling, I cut it and tied the end to a substantial piece of wreckage.

Just at that moment McIlwee arrived. I could tell by the sour expression on his face that he suspected at once what we had done. I suppose he was vexed because he could not now add another notch to his body count. I made the situation clear with explanatory gestures. He and his buddy ascended the line to take charge of surface operations. Relieved of further obligation, Messina and I continued with our original plan to recover the capstan. By the time we got back to the boat with our prize in tow, the body had already been retrieved. Decomposition and the stench of rotting flesh had made the surface recovery a nauseating job. Several divers were sickened by the episode. They puked over the gunwale despite the flat state of the sea. During the ride back we avoided the port side of the boat where the body lay covered with a tarp. The coroner met us at the dock to take possession.

The diver's grapevine is as malicious and misinformed as every other gossip line. After multiple retellings, the story that circled back to me was at variance with the truth on a number of points. The major misconceptions were that the sole purpose of the charter was to recover the body (instead of a normal club charter to which such a mission was attached), and that after recovering the body a *large* artifact was recovered. The latter statement was true, but disapproval was based upon the falsehood that the recovery of the body was a specific objective for which we all volunteered, and that under the circumstances we had no right to recover artifacts as well.

Additional censure came from the fact that we recovered the capstan *on the same dive*, the implication being that had we surfaced and gone back down, the interval would have demonstrated sufficient respect for the dead to be socially acceptable. I fail to fathom this mode of thinking. Splitting the discrete quantum of a dive into separate components is an expedient of irrational thought. And what the size of the artifact had to do with anything is beyond all comprehension.

Reverence for corporeal sanctity is a belief that not everyone shares. I regard the body as the carrier of the mind, of the personality, or, if you prefer, of the soul. From this viewpoint, the body is equivalent to an automobile whose purpose is to convey a person through life. Once life has fled - that is, when the chemical interactions that support the mind are interrupted - the persona ceases to exist and the deceased body has no further purpose. I conceptualize a person not as a physical body but as a consciousness whose nature is ethereal and ineffable. As much as I loved my grandfather I have never visited his grave because he is not there. Only the vehicle that carried him through life is buried there. He is now a memory which I carry with me always.

Bent Again

Enough philosophical musing for the nonce. The last EDA charter of the year occurred on December 29. I dived with Tom Roach to the

Balaena, a wooden-hulled sailing ship that lies 170 feet deep on the edge of the Mud Hole. Twelve minutes into the dive I felt my air running out. I'm pretty bad on air, but not that bad. Yet the gauge that showed my tanks full on the boat now showed that they were empty. I switched to my pony bottle and signaled to Roach that I was leaving. He continued his dive while I ascended quickly but breathing normally.

Back at the boat I paused halfway up the ladder. When I explained the reason for my early return, and my misapprehension, Hoffman leaned over the rail and checked my tank valves. There came a loud hiss as air cascaded rapidly from one tank to the other. I was wearing two singles whose valves were connected by a crossover manifold, called a cheater bar, whose single orifice fed the air from both tanks to my primary regulator. I had not opened one of the valves. As soon as the tanks equalized and my gauge showed them to be half full, I slid off the ladder and went back down to the wreck. I spent six minutes in the vicinity of the grapnel, then made a normal ascent and decompressed for five minutes - all the time that was required by my meter.

Afterward, I doffed my tanks and sat in the sun on the back of the boat trying to regain some warmth. John Pletnik started a conversation. He had once had what he thought might have been the bends, and wanted to know what it was like when I felt numbness in my extremities several weeks earlier after surveying the helm on the *Ioannis P. Goulandris*. I explained how my right arm and leg had tingled eerily as if the circulation had been cut off and they had "gone to sleep," and how the skin on the right side of my chest and abdomen felt strangely anesthetized, the way gums and lips feel after being injected with Novocain.

"That's odd," I said. "It feels the same way now." I rubbed my side through the material of the drysuit. I thought my skin was numb from cold, but suddenly I lost my balance and pitched forward into Pletnik's arms. He let me down to the deck, rolled me over onto my back, and yelled for help. Hoffman was there in an instant. I didn't lose consciousness, but I soon ascertained that I had lost all feeling on the right side of my body. I couldn't move my arm or leg at all, the affected parts were totally insensitive to touch, and my head was assaulted by waves of dizziness.

Hoffman asked what I wanted to do. Although it was difficult to think the situation through, it was obvious to me that I couldn't go back in the water as I had the year before. I was dragged out of the way so that people returning from their dives wouldn't trip over me. Everyone stooped down to commiserate with me, but no one could say with any authority what was wrong and no one had any suggestions to make. Hoffman asked if I wanted him to call the Coast Guard.

1970's: Decompression Comes of Age - 245

Nowadays a dive boat captain would not hesitate to radio a report, nor would he ask for a patient's recommendation. He would take control of the situation. Or, if he were not a diver himself and not conversant with diving procedure, the dive master would take control. Roach was back on board, but he didn't understand any more than anyone else about the esoteric discipline of decompression injuries. In those days we were all in the same boat, so to speak, in our ignorance. So the decision on how to proceed was left completely in my hands - or in the hand that still had feeling, and in the part of my mind that could still think clearly.

When my level of awareness returned near to normal, and the degree of paralysis became evident to me, I asked Hoffman to arrange for helicopter evacuation to a recompression chamber. Feelings of deja vu washed over me: once again I lay helpless on my back staring up at the bright blue sky, seriously injured and unable to move, waiting to be whisked away by helicopter in order to obtain medical attention. Only this time my right side was affected instead of my left.

A helicopter pickup at sea was a big event. I had seen it only once: the previous May when Danny Bressette suffered intense abdominal pain after a dive to 130 feet. People on the boat were as flushed with excitement as they were with concern. Now Bressette crouched by my side to reassure me about the efficacy of recompression treatment.

In the meantime, I began to regain the use of my fingers. I kept flexing my fist, thinking that the increase in circulation would help. Soon I was able to move my leg a bit. Bressette pinched my toes periodically to test my feeling and reflexes. Eventually, I was able to sit up. No longer did the gunwales protect me from the bitterly cold breeze. Pletnik half-carried me into the engine room for warmth.

As the paralysis gradually diminished, Pletnik helped me to get out of my drysuit and into regular clothes: quite an ordeal in the close confines available behind the engine. By the time I heard the blades of the approaching helicopter slicing through the air, I was fully dressed in longjohns, boots, gloves, trousers, flannel shirt, sweater, and hooded parka. And I could stand up, albeit shakily, on my own.

The *Sea Lion* maintained steady speed in the direction advised by the Coast Guard pilot, with whom Hoffman was in constant radio contact. Customers and crew cleared a space on the deck and either packed away loose equipment or lashed it down so it wouldn't get blown overboard by the downdraft. The helicopter approached from the stern, presaged by the raucous noise of its engine and the wind generated by its swiftly rotating blades. Its position with respect to the water was evinced by a concentric ring of spreading ripples. Someone brought me my wallet, but my car keys were entrusted to Frank Messina, who lived near me and Roach and who later drove my car to Roach's house.

246 - The *Lusitania* Controversies

A wire-mesh basket was lowered from the cargo bay on a stainless steel cable. Everyone with a camera got in position to record the event. By this time I had recovered almost full use of my limbs and faculties, so I got someone to get my camera for me and hang it about my neck. The basket touched the rail - we had been warned not to touch it till the static charge was grounded by the boat - bounced up, then came back down within reach as the pilot jockeyed the helicopter for position and the crew chief drew out the cable to the basket. As soon as the basket touched down for the second time, someone unclipped it, signaled to the pilot, and the helicopter veered off to stand by until I was ready for the lift. I climbed into the basket. Roach shoved a sheet of paper into my pocket in case I passed out. Written in ink were my bottom times and abbreviated surface interval.

Now came the tricky part. Too well I remembered the story told by Don Nitsch about a pickup that went awry. On that occasion, after lifting an unconscious diver off the deck in a stretcher, the helicopter lost altitude when wind sheer swept it sideways. The stretcher splashed into the water and the patient was submerged. The helicopter then proceeded to bob up and down as the crew chief frantically reeled in the cable, with the result that the stretcher and the patient who was strapped to it were alternately dragged and dropped across the surface in what could have been a comic scene from the Keystone Kops had the plight not been so perilous. Nitsch and another person leaped overboard and swam to the rescue. Just as they reached the patient in the stretcher, the helicopter settled down practically on top of them. As the heavy machine hit the sea, kept upright only by upward force generated by the whirling blades, the cargo deck sank underwater, and the patient and both swimmers were washed into the compartment by the flow. The pilot applied emergency power to the rotor. The helicopter groaned under the excessive weight as it slowly rose above the waves with hundreds of gallons of water gushing out of the cargo bay. Nitsch and his companion got a ride to the hospital, where they had to stand around in wetsuits till the boat reached port and someone brought them their clothes. The patient also survived.

I decided that I didn't want to be strapped in, just in case I had to make a fast getaway. The helicopter circled around the boat, approached with the cable hanging just above the level of the rail, then hovered directly overhead as the basket was secured to the clip. The mighty machine then backed and lifted so smoothly that I never felt a lurch. I snapped a sequence of pictures from the air - my motto as a photographer has always been the same: no matter what happens, get the shot. The basket was hauled up to the cargo bay, the davit was swung inboard, and I climbed out onto the deck. The engine noise was so incredibly loud that I couldn't hear myself think. The crew chief signaled for me to secure myself to a safety harness.

On the way to Mount Sinai Hospital the police were notified to prepare a place to land. They were still running kids off the ball field when the airborne ambulance arrived. I was sorry for interrupting their game. By this time I felt fully recovered, with not a hint of numbness anywhere and with complete mobility in my arms and legs. That was why, when the helicopter landed and the crew chief slid back the door, I jumped out onto the grass on my own and struck out jauntily to where a medical response team stood by a stretcher on wheels.

I walked toward them as they hurried toward me. Because of my healthy gait and the surplus Air Force parka I wore, they mistook me for a helicopter crew member and brushed right by. We could have been a couple of star-crossed lovers whose outreaching arms somehow failed to connect. I felt foolish standing alone in the outfield as the orderlies peered into the cargo bay for a patient who was supposed to be paralyzed. The crew chief pointed in my direction. This time I stood still as they approached. I explained that I was the patient and that I was feeling much better. Nevertheless, they insisted that I lie on the stretcher so they could roll me to the chamber. They strapped me in for the high-speed dash.

We bounced and clattered past the bases like a hitter trying to outrun the ball. It was smoother on the macadam. All the time, a medical assistant ran alongside asking me questions and jotting notes on a clipboard. We pushed through the doors into a long corridor which, unfortunately for the orderly, soon narrowed from three lanes to two. The last I saw of him, he was wiped out at high speed by a partition that suddenly jutted out from the wall.

The doctor didn't seem to know very much about recompression treatment. The chamber was used primarily to administer hyperbaric oxygen to patients with infections that refused to heal. He pulled out a U.S. Navy diving manual and looked up Table 5. Despite the fact that I felt completely cured, he insisted upon a preventative treatment of two and a half hours. Accompanying me were a technician and a patient with gangrene. This time I asked for extra blankets.

I couldn't report feeling relief at 60 feet. Nor did I feel any different after treatment. I stayed in the hospital overnight for observation, and the next day, after a superficial examination, the doctor released me from his care. Getting home was an adventure. First the bus driver threw me off the bus because I didn't have the exact change to pay the fare. I explained that I had just been released from the hospital and that was the only money I had, but this was New York City so compassion was not forthcoming. Nor would any clerks in the nearby stores change a dollar. Eventually I had to buy something in order to obtain the needed coinage.

The train station was a nightmare montage of hustling, bustling people, none of whom would give me the time of day let alone direct me through the maze of underground passageways. The manmade cavern was inhabited by a menagerie of diverse types: beggars and businessmen, drunks and dowagers, hippies and harlots, spaced-out teenagers, worldly-wise elderly, the wild, the phlegmatic, the flamboyant, the indifferent, the unconscious, and, by my presence, the gauche. It was a place of juxtapositions where abnormality was the norm. The weird dress and strange manner were shouts for attention, pleas for notice, struggles for consideration, all manifested out of the psychological need to express individuality in a world otherwise subordinated by the sheer numbers in crowd. Such seemed New York City to an unsophisticate yokel.

Somehow I tracked down the train to Philadelphia. There I boarded another train and then got on a bus that dropped me off within walking distance of Roach's residence, where my car was parked. As a result of my escapade - for forgetting to open one of my tanks before the dive - Roach awarded me the "tourist" diver prize at Hoffman's New Year's Eve party the following night: a tee shirt with the word TOURIST printed on the front in boldface capitals. I donned it with great humility in front of my fellow divers.

If you screwed up with this outfit, no one ever let you forget it. I went diving the very next week. I wore the tee shirt on the boat with pride.

Tragedy Close to Home

I continued to dive with EDA throughout the winter. Many of the dives were shallow, but trips to the *Ayuruoca* kept us in trim for a special deep dive to come. I tried to keep my bottom times short in order to curtail decompression in the frigid water. My longest hang was in April - thirty-five minutes - and even in a drysuit I was chilled by the 46° water. But my enthusiasm couldn't be dampened by cold. I was full of anticipation for the trip that was scheduled in June.

An interlude with tragedy occurred in May. We sat around the deck of the *Sea Lion* after a morning dive to the *Pinta*. The bite in the air was more than offset by the warming rays of the sun. The only one in the water was the mate, John Pletnik, who seemed to be making an unusually long dive. Long grew longer, and still no bubbles appeared on the anchor line where he should have been decompressing. The assumption made in such circumstances was that the missing diver must be decompressing remotely on his own sisal line. After a while, Hoffman asked if anyone knew precisely when Pletnik went down. In those days we didn't keep a log for entries and exits. Everyone was pretty much on his own. By best estimate nearly an hour had passed since his descent.

At this point we began to scan the surface for bubbles in the dis-

1970's: Decompression Comes of Age - 249

tance. I was eating a sandwich at the time, but as the minutes crawled by I slowly lost my appetite until I could no longer swallow. It is amazing how long the human mind can refuse to accept a truth. We began to make excuses for Pletnik's overdue return: overfilled tanks, his low breathing rate, his remarkable breath-holding capacity, wavelets obscuring his bubbles, surfacing downcurrent out of sight. As the hour became an hour and a half we began to experience doubt. Every minute that passed lessened his chances for survival. Every second became an exercise in agony.

We were a boatload of isolated minds attempting to communicate hope, and by doing so, to turn that hope into reality. Each of us eventually admitted the unwanted certainty according to his own needs and preconceptions. The last convert yielded after two hours of anxiety had passed. We were all believers. John Pletnik was dead.

Nor did we find his body on the second dive. Teams of divers fanned out around the wreck in a broad search pattern that encompassed the exterior hull and surrounding debris field. I entered the capacious holds full of lumber as Roach remained by the hatch coamings and guided my progress with his light, which doubled as a beacon that shone the way to safety. I moved cautiously from one hold to the other, poking my light into the deepest cavities between boards that were twenty feet long, while trying not to dislodge the timbers that were stacked haphazardly like a child's pile of pickup sticks.

Hoffman reported the incident by radio, so the police were waiting for us at the dock. Each of us was questioned in what was for the police a homicide investigation. They did not assume that the death was accidental until they had the facts in hand. Then Hoffman undertook the sad task of calling Pletnik's girlfriend to relay the bad news. She was hysterical.

Two days later, on Tuesday, those of us who could get off work went out again on the *Sea Lion* on a body recovery mission. Along to lead the effort was Little John's best friend, Big John Dudas. Because of bad feelings provoked by past altercations over the salvage of certain prime artifacts, Dudas no longer dived with Hoffman or with EDA. This was the first time in years that he had stepped foot on the *Sea Lion's* deck. Which goes to show how petty differences created by people in their greed can be surmounted by events of greater calling.

Our search plan was based upon the assumption that Pletnik was lost inside. Rather than have everyone enter the water at once and shotgun the wreck in the blind, we planned to dive on a rotational basis so that each team could communicate its findings to the next, which would then continue the search from where the previous team left off. Because Dudas knew the wreck better than anyone, he was chosen to lead the first team. I was selected to lead the second.

Since I had already checked the holds with Roach, Dudas and

Bressette began their search in the stern superstructure and worked their way down to the lower decks. They returned in an incredibly short time with the information that they had found the body in the after steerage compartment. I shivered involuntarily because there was only one way to reach the after steerage, and that was through the central passageway in which I had gotten lost five months earlier after recovering the helm from the *Ioannis P. Goulandris*. They weren't able to pull the body out of the room because it was stuck, and in trying to loosen it they had stirred up the sediment so much that they were no longer able to see. They secured a line to the tank manifold and fed it through the intersecting corridors to the outer doorway. They recovered Pletnik's light, whose battery was fully discharged. Pletnik carried no back-up. Dudas and Bressette concluded, and all concurred, that Pletnik's death was brought about by the failure of his light. Plunged into darkness, he was unable to find his way out of the narrow crawl space before his air ran out.

The job of bringing out the body now devolved upon me. For this complex task we formed a team of four. Bob Archambault went inside with me. Roach and Nitsch tended the line from outside. I entered first with Archambault on my heels. I followed the thick white nylon rope through the maze of twists and turns, left Archambault at the bottom of the horizontal shaft, then continued on alone and emerged in short order in the central passageway. From there I shone my light through the narrow hatchway into the after steerage compartment. The body lay supine about twenty feet inside the room, head foremost. The arms were bent upward in a pose of supplication induced by rigor mortis. I tugged on the rope, hoping to bring the body toward me without having to go in any farther, but it didn't budge.

It took great effort of will to venture beyond the point at which I had once become disoriented so frightfully. Only the presence of the guideline to the outside gave me the timid confidence to go where I inwardly feared to tread. I squeezed through the opening and slowly worked my way along the corridor to the body, careful to avoid kicking up loose silt by pulling myself forward with my hands on the exposed iron beams.

Pletnik looked up with an expression of pain and fear through a mask that was unflooded. I saw in his face the abject terror of those final desperate moments. There is no pain in death, for in death comes surcease of all sensation. It is the *fear* of death that is so awful. Thus fear equates to emotional pain and death becomes irreversible release. The vital essence that was Pletnik no longer existed. Yet I couldn't help but feel anguish over the pain and fear he must have felt in his final moments - more so than I felt for Gary Ford, who was a stranger to me. Pletnik was my friend.

I forced these musings out of my mind. Ignoring Pletnik's face, I squirmed over his body with my tanks banging against the metal overhead, got past him, then drew in my legs so I could turn around and face the way out of this cramped, closet-sized compartment. I shoved the corpse back and forth as I examined the area carefully for snags. Right away I saw that Pletnik's weightbelt was the cause of the problem. The weights were hung up on an exposed iron beam. When I flipped open the buckle, the belt fell away and the corpse instantly gained enough buoyancy to rise slightly off the bulkhead framework. Now the body moved unencumbered.

I pulled myself forward till we were face to face and I was positioned within the open circle of the arms in rigor. As I drew the body along, Archambault, unseen, hauled in the slack and kept the line taut. I rolled sideways as we approached the hatchway so we could slide through. Pletnik and I pirouetted like two ballerinas in a ghastly danse macabre. My pulling combined with Archambault's hauling imparted considerable momentum. The corpse and I charged through the hatchway. But I hadn't calculated on the funneling effect of the narrow opening, which was not wide enough for the two of us to pass through together. The curved tops of our tanks acted as wedges, and in an instant we were thrust together so tight that our chests bonded like two laminations of plywood, and my mask was jammed hard against Pletnik's. Once again I found myself held in Pletnik's arms. We were stuck.

I cannot adequately describe my terror. I tried to break free, but I couldn't back up because Archambault was maintaining such a determined strain on the line. Frantically I yanked on the rope as a signal for slack. Archambault let go. By pushing hard against the transverse bulkhead with my free hand I managed to unwedge myself and break loose from Pletnik's embrace. I was breathing pretty hard from the strain - emotional as well as physical. I couldn't get out through the opening as long as it was blocked by the corpse, so I heaved it back into the room before Archambault again took up the slack. I wanted out of there in the worst way.

As I charged past the body I was brought to a sudden halt. I struggled, but I couldn't move forward. Something was holding me back. I swung my light in the close confines of the room, now silting heavily, and what I saw in the diminishing visibility added immeasurably to my fright. *The handle of my spare light was locked in the grip of Pletnik's frozen fist.*

My imagination was instantly fired by superstitious implication, and I completely lost control of my equanimity. I yanked at the light but it would not come free. I had to ease the handle out through semi-closed fingers. Then I kicked away from the body in sheer panic, banged my pony bottle against the metal lip of the doorway, backed up and pulled myself through, and erupted into the central passageway with my heart pounding like a kettle drum beating the call to battle.

The accidental placement of my light in Pletnik's hand had wholly unnerved me. I felt the blood pulsing against my temples, threatening to burst an artery, and I was shaking with nervous alarm. I gripped the steel bulkhead to steady myself while I sought to regain my composure. In my current state of demoralization, I felt too incapacitated to complete the mission. Yet a part of me did not want to fail in a task that I was physically able to do. My rational being fought to keep my irrational fears in check. I kept reminding myself that I was safe, that there was nothing threatening me but artificial constructs given credence by vivid imagination. I refused to subordinate my reason to my id.

It took at least a minute or two for me to control my breathing and timidity. During that time I hovered in virtual darkness within reach of a corpse that I intellectualized couldn't hurt me. Archambault, too, lay somewhere in the darkness, waiting patiently and wondering what was occurring deeper in the bowels of the wreck. He pulled the line taut. I calmed myself down, forced my upper body through the doorway, grabbed Pletnik's manifold, and gently pulled him out of his temporary tomb. I could not bend the arms because of their stiffness.

I fed the body down to the bottom of the room and guided it toward the opening that was now enshrouded with silt. Archambault maintained tension on the line. I worked Pletnik's head and shoulders into the shaft by feel. Archambault pulled while I pushed. The corpse went in easily at first, full length, then jammed. The tanks and the rigidly outthrust arms created an eccentric shape that snagged on the partially collapsed interior of the shaft and hung up on exposed beams and fallen jagged debris, and that no amount of tugging back and forth would release.

Now I was trapped again and more terrified than ever. I grabbed Pletnik's legs and tried to pull the body back out of the shaft; no go. How long this tug of war continued I have no idea, but undoubtedly not nearly as long as it felt at the time. I jumped to the conclusion that the body was permanently stuck - or that it would be stuck for a length of time that exceeded how long I could survive on the air remaining in my tanks.

In frantic desperation I abandoned the body and made an irreversible leap of faith. I rose above the silt to the top of the passageway and worked my way forward. I passed the rooms I had briefly explored before, and quickly found myself in unfamiliar territory. My only comfort was the clarity of the water ahead of me. The corridor seemed endless. In fact, it was probably less than fifty feet. But crawling fifty feet through unknown darkness, expecting a literal dead end to appear, made it seem like more than a mile.

And then there was light! Only a faint distant glow at first, but it brightened my heart and illuminated my prospects like a sunbeam after a long Arctic night. Then I entered a transverse corridor that extended across the after bulkhead of the first cargo hold. A moment later I spilled out of the wreck into glorious open water.

Immediately, I raced along the upper hull to the open doorway I had originally entered. Roach and Nitsch didn't know who I was when I shoved them out of the way. Archambault did a double take. Somehow he had managed to free the body and drag it through the connecting corridors alone. When I appeared by his side on top of the wreck he was totally nonplused, since he thought I was at the other end of the line pushing the corpse. He kept jerking frantically on the rope but the body was stuck again. I could plainly see that the back of the head was being slammed against a steel beam. I signaled for Archambault to halt while I reached inside the doorway and maneuvered the body into the open.

Roach and Nitsch took over once the corpse was outside the wreck. They carried it up the line while Archambault and I decompressed. By the time we got back aboard the boat, the gruesome task of retrieval was done. As our resident doctor, Gene DellaBadia conducted a perfunctory examination of the body. The coroner met us at the dock to take possession.

Pletnik was dead and there was no bringing him back. But there was much to be learned by his untimely demise. In most cases of this nature, the cause of death is given as drowning or suffocation. These medical distinctions are employed to fill out death certificates and create diving fatality statistics, but they yield no useful information. Noting that a person inhaled water into the lungs (drowning) or did not inhale water (suffocation) describes only that person's final breath. It doesn't tell us anything about the *reason* for the medical cause of death. It's like claiming that someone found deceased on the sidewalk died from multiple fractures and internal contusions. Did the decedent's parachute fail to open, or did someone shove him out of a tenth story window? These, I contend, are what we need to know.

In the context of diving, we can go a step beyond the medical cause and say that a person ran out of air, but even that explanation is unsatisfactory. We need to know *why* he ran out of air. Most diving accidents can be grouped under three major categories: poor planning, poor execution, or overcome by unforeseen events.

Gary Ford died from poor execution because he failed to leave the bottom with sufficient air in reserve. John Pletnik died from poor planning because he carried only one light under circumstances that reason tells us required at least two. I knew this because of my extensive spelunking background. The rule of thumb in caving is to carry at least three sources of illumination (carbide lamp, flashlight, and candles). But what seems obvious today was not necessarily manifest at the time, when wreck penetration was a new and poorly understood activity without written guidelines or established protocols.

Pletnik mistakenly trusted the reliability of his light, which had never failed him before. Rechargeable batteries were a recent innovation. None of us fully appreciated the way nickel-cadmium batteries burned out so rapidly at the end of their cycle, compared to non-rechargeable dry cells in common usage, which faded gradually. Thus when a bulb dimmed, there might not be enough power left in the battery pack for a person to make his exit before the light went out altogether. We speculated endlessly on whether Pletnik charged his light sufficiently or at all after the previous day's dives. No one knew for sure. But the fact remained that a combination of events that were potentially within his control precipitated in his demise.

Diver error is the most common cause of death in wreck-diving. That some of us committed such errors but were lucky enough to survive them casts no aspersions on those who paid the ultimate price for their mistakes.

I went diving again two days later. This is not to imply that I went unaffected by the dreadful matter of Pletnik's death and body recovery. Quite the contrary, I was plagued by disquieting momentary images for six months afterward, during which time I returned to a state of mind last suffered in my pre-teen years.

Flashback

Flashbacks were nothing new to me. I began having them upon my return to pain-reduced consciousness several months after being wounded in Vietnam. These flashbacks usually took the form of instantaneous visualizations of my final observation of the exploding muzzle flash and the kneeling enemy soldier behind the gun: a traumatic experience that was replayed over and over in my mind. I should be quick to point out that these flashbacks were seldom debilitating.

A flashback lasting a fraction of a second might occur on a crowded sidewalk while I was talking with a co-worker on a sunny afternoon. I never missed a step or stumbled over a word, and my companion wouldn't know that a ghastly mental picture had erupted inside my head. I ignored the fleeting memory and quickly forgot that it occurred. After some five years these daytime flashbacks ceased.

Nighttime flashbacks continued on a nightly basis for the next ten years or so, generally as I lay in bed about to fall asleep, and only when I slept alone. During the subsequent decade the frequency of occurrence diminished to several times a week. Now I have them only once or twice a month, but the duration has increased, sometimes lasting an hour or more as every agonizing detail of that near-death experience is re-enacted while I toss and turn on my pillow.

Also, occasionally, I've had freeze-frame episodes that stopped me in my tracks. These have been induced by the sound of helicopter blades, particularly hueys. During these long-lasting, lucid retrospects, my body went rigid, my eyes glazed, a discordant chill coursed along my spine, and a storyboard from the battlefield engulfed my mind. Outwardly, I've been told, I appeared mesmerized. But psychically I was flushed with excitement: an evocation of the thrill and sheer vitality I tasted during firefights.

Don't jump to the conclusion that this uncontrolled response to stimulus symbolizes a subconscious yearning for bloodlust. It does not. War and its combatant components (charges, skirmishes, firefights, and battles) are filled with horror and the fear of death or dismemberment. A healthy mind does not dwell in distress, but forgets or represses pain while it reinforces therapeutic meditation. After time, the unwholesome feelings of fear and horror fade, leaving conspicuous those emotions which were subjugated by more powerful primal influences. On a superficial level, without reflection or deep introspection, I was responding to that aspect of battle that inspires the most stimulating emotion one can ever experience: the fierce exhilaration of surviving a life-threatening ordeal. The euphoria of life is never so intense as it is in the moment after a close call with death.

Only one who has been there can truly understand. You might just as well describe color to the blind or sound to the deaf.

The sense of near-death euphoria is not a feeling that a rational person seeks. A person with a death wish has lost control of his reason, or suffers from cathexis. In a similar vein, one who is fixated on a past powerful experience is caught in a mental feedback loop from which he is unable to escape. Veterans who thrive on retelling war stories are trying subconsciously to recapture those moments of excitement which life in civilization cannot provide.

In addition to this legacy of combat, I began to see Pletnik's pleading face in the shadows. Once again I became afraid of the dark, fearful of partially closed closet doors, spooked by nighttime shadows in my own familiar house. It was like returning to childhood uncertainty. When I closed the closet door in my bedroom I had to make sure that the latch clicked aloud, or I could not get to sleep, fearing that otherwise a specter of the dead would push the door open during the night and slip unbidden into the room. Opening the door in the pre-dawn hours as I dressed for work required uncommon fortitude, as I expected some dreadful incarnation to leap out at me. I did not have nightmares. These apparitions were expressions of a fully conscious mind.

Over the months these torments tapered off and my night life returned to normal.

Andrea Doria Figment

The non-event of 1973 was a trip to the *Andrea Doria* that didn't come off. Roach chartered a commercial vessel that was supposed to pick us up in Brielle. In the days preceding the embarkation date he wasn't able to contact the owners by phone in order to pinpoint the time of departure, so we all arrived bright and early Monday morning and waited for the boat to appear. It never did. Our disappointment was immeasurable. No one blamed Roach for the snafu, but I'm sure that as the organizer he felt personally accountable for nonfeasance. Not until Wednesday did the irresponsible owners deem to call and mention that the boat was docked hundreds of miles away with mechanical difficulties. They scoffed as if it were of no importance that a dozen people were waiting anxiously for word of the boat's arrival, and that these people had given up a week's wages to be there.

The Figment Materializes

For a year Roach suffered jibes and disrepute for the boat's non-performance, so it was with some trepidation that he scheduled another trip to the *Andrea Doria* for 1974. He need not have worried, however, for this trip came to pass without a hitch. Transportation to the site was provided by the research vessel *Atlantic Twin*, a 90-foot diesel-driven catamaran with a crew of four: captain, mate, engineer, and cook. This was the same vessel that took Alan Krasberg's expedition to the wreck in 1968. For wreck-divers used to cramped and Spartan accommodations on what are commonly called "cattle boats," the *Atlantic Twin* was sheer luxury. The interior was air-conditioned throughout, bunks were provided for ten of the twelve participants (two people slept on benches), meals were prepared in an ample galley and eaten at built-in tables, and the crew did all the work.

The *Atlantic Twin* was no speed-boat: her barnacle-encrusted hull made barely eight knots through the perfectly flat seas that blessed the entire trip. It took twenty-four hours to get to the wreck from Brielle, a distance of 190 nautical miles. The captain ran his ship with calm professionalism, much different from the seat-of-the-pants operations we were used to.

None of us had been to the *Andrea Doria* before. We divided ourselves into six teams. Tom Roach dived with Danny Bressette; Jan Nagrowski paired up with Ron Burdewick, part owner of the *Sea Hunter*; Bob Archambault went in with John Asqui; Joel Entler, a school teacher, teamed up with Ray Bailey, a bookie and bar owner; fireman Donn Dwyer entered the water at the same time as trucker Jim Snyder, after which each went his separate way.

My buddy was John Starace, owner of a shooting range. Because Starace was the oldest diver on the trip and I was the youngest, we were dubbed "the Ford and the Ferrari." It was hoped that his mellowed maturity would temper my youthful vigor and exuberance, and that my strength and stamina would overcome his waning robustness. We were perfectly matched.

After the previous year's failed expectation of the trip to materialize, I was particularly eager to touch down on the haunted, hallowed hull. In those days, to dive the *Andrea Doria* was to achieve immortality among the pantheon of prominent wreck-divers. Only a handful of people had made this underwater rite of passage. To merely touch the upper rail was like sliding into home plate after a long hit to center field, with the umpire yelling "Safe!" It was a home run that could never be disputed.

My agenda went beyond such a nebulous touch of fame. In my mesh bag I carried a large liftbag and a specially designed cable leader small enough to weave through a bolt hole. I was going after the brass builder's plaque mounted outside the forward bulkhead of the wheelhouse. I had studied the plans and photographs carefully, knew exactly where to go to find the prize, and how to recognize the proper level by the shape and number of windows that pierced the steel. I planned to slice through one bolt with a hammer and chisel, secure the liftbag with the cable leader, partially inflate the liftbag, then pry off the heavy plaque with a crowbar or chisel off the other three bolts. Rumor that circulated from a salvage expedition the year before, under the aegis of Don Rodocker and Chris DeLucchi, was that the top deck of the wheelhouse had collapsed. I found this difficult to accept.

Roach and Bressette were first down the anchor line. Their job was to move the grapnel to a location suitable for a permanent moor, away from upreaching obstructions that might snag the mooring line as the boat circled the wreck with the changing tide. Starace and I followed ten minutes later. I dragged the permanent line by the chain and shackle against a barely perceptible current. Starace was close on my heels, hauling the rope. When I alighted on the forward hull I exhibited no outward sign of the way I thrilled inside from touching the silent Siren that had beckoned to me for a year. I went about my assigned task, secure in the knowledge that, if all else failed, I had already achieved a major goal just by being there.

Roach and Bressette were waiting for us on the high side of the wreck. I swam past them looking for a place to wrap the chain, found a scupper on the well deck immediately in front of the bridge - the perfect place to be for my chosen objective - and slid the shackle and chain through the opening. Roach went underneath, grabbed the shackle, and hefted it up and around the outside of the scupper. Together we secured it.

While we were doing this, Nagrowski and Burdewick arrived five minutes ahead of schedule. Seeing that we were in control of the situ-

258 - The *Lusitania* Controversies

ation, they kept right on going, heading aft. That was how they happened to be the first to come across the loose porthole lying atop the hull, near the Foyer Deck entry doors, where Rodocker and DeLucchi had left it after torching it out of one of the doors. The porthole had been so unimportant to them that they left it behind. Portholes are meaningless outside the wreck diving realm - they are equivalent to wooden window frames salvaged from an abandoned farm house. Burdewick attached a liftbag to one of the dogs, but did not bother to send it to the surface, saving it for another dive.

Meanwhile, with ambient light visibility exceeding thirty feet, I oriented myself on the bridgewing and descended to the ship's centerline where the builder's plaque was mounted. Starace remained above and marked the spot with his light. Then I discovered the awful fact that the rumor reported the truth: the top deck of the wheelhouse, from which Dudas had recovered the magnetic compass, was gone - and so was the bulkhead to which the builder's plaque was bolted. I peered in through broken glass at a room with no overhead - the overhead that should have been the deck of the wheelhouse. My hopes for a unique artifact were dashed!

It had taken me three years to surpass the personal depth record that I once set on a Bahamian reef. This time I was properly equipped - instead of a single tank I wore doubles and a pony bottle. And I was fully prepared - in mind as well in gear - to handle remote decompression should the necessity arise.

On the second dive, that afternoon, we did a little exploring. Nothing significant occurred, other than I noticed that Burdewick still had not sent up his porthole. He felt no pressing need to do so because it was appropriately tagged, and because the mutual trust among team members gave him the confidence to expect that the artifact would be left alone. Starace and I delved into the Promenade Deck, where rectangular brass windows had fallen out of their frames, but we made no recoveries on that dive because we ran short of time.

That night a fog developed so thick that from the middle of the *Atlantic Twin* I couldn't see either end. Visibility on deck was about twenty feet. When the cabin door was opened, thick white billows rolled inside like water from a sluice gate in slow motion. The food was excellent, the coffee hot, the crew cheerful, and the camaraderie at its peak - perhaps extreme.

For example, as Dwyer was climbing up the ladder he asked Roach to take his picture with Dwyer's camera. Roach complied with exasperation because Dwyer was always asking to have pictures taken of himself. Later, when Dwyer was underwater, Bressette dropped his drawers and let Roach photograph his genitals in extreme close-up, hoping to surprise Dwyer when he had the film developed. A week or so later Dwyer was suitably surprised - and so were his wife and kids.

We were all in high spirits. Just the fact that we had dived the wreck made the expedition a success. And most of us had hopes of retrieving a window on the morrow. A souvenir was a bonus to some, the end-all to others, depending upon one's frame of mind.

Starace and I hit the water early. I wanted a window badly, but salvage was only part of the full-scale dive I had planned. We swam past Burdewick's porthole, dropped into the Promenade Deck, and settled down next to a pile of windows. Each window had a handle conveniently sized so a liftbag could be attached with ease. Starace unrolled the liftbag while I secured the leader. Then, as I pulled the window up out of the debris, he inflated the liftbag until the window began to rise. The liftbag passed through the broad opening in the framework, but the window snagged temporarily on the lip. I freed the window and watched it soar to the surface untethered. The mate was waiting in the chase boat to recover the many windows that the group was planning to recover.

We rose out of the Promenade Deck and swam along the hull to the Foyer Deck, two decks lower, for phase two of our dive. We hovered above the door from which Burdewick's porthole had been torched. The opening was about two and a half feet square: jagged, dark, and foreboding. My beam failed to reach the bottom of the seemingly endless abyss. Only Rodocker and DeLucchi had entered the blackness beyond, and they were tethered to umbilical hoses when they did so. I exchanged okay signals with Starace, positioned myself by standing vertically over the hole, let air out of my suit, then squirmed and scraped through the aperture.

It was like dropping into a tight submarine hatch except that the hole was square instead of round. I had to fold my arms above my head in order to fit. Inside, I found myself in a corner bounded by two steel plates, one of which was the deck, smooth and unobstructed, and the other a partition. The partition ended a few feet down, admitting me into a vast open gallery whose only boundary was the ceiling eight feet away. All-consuming blackness extended down and to both sides as far as my light would shine.

Starace's light shone like a beacon above me, but as I descended deeper into the open interior, my exhaust bubbles created a barrier that blocked all illumination from above. The ceiling was a disquieting mass of loose wires and dangling electrical cables that stretched out like a talus slope the farther down I went. In the Stygian blackness, my sole means of orientation was my depth gauge, which I monitored constantly.

I hoped to reach the debris field from which Rodocker and DeLucchi had recovered a few souvenir items. I kept falling, like a skydiver in extreme slow motion, peering beneath me for signs of the bottom. As I descended, I became aware, peripherally at first, of the ceiling closing in on me from the right. At a depth of 190 feet the passage shrank to half its original width. I could no longer ignore the encroaching tangle of wires and cables. They appeared to be reaching out for me like a many-tentacled beast. I shoved them away with my hand until the gap closed in to the danger point.

At 200 feet I rolled into an upright posture and pressed my drysuit's inflation valve. Air hissed into the suit, but not fast enough to offset my downward momentum. I sank with what seemed like accelerating speed - the same sensation one has prior to an automobile collision - with events telescoping out of control. I kicked hard. My fins beat against steel wires that once supported the ceiling panels, and against copper cables that were thickly insulated for high voltage transmission or multi-wrapped for telephone communication. The room was a snake pit full of darting steel necks that were as deadly as their biological counterpart.

I dragged my fingers down the silt-covered deck, wishing I had talons instead of rubber-covered fingertips. When my right elbow brushed past the impenetrable metal mesh, I pulled my arms in close. I was sliding down a funnel that pinched together at the bottom like a Chinese handcuff. The last time I looked at my depth gauge it read 205 feet. Now I could no longer concern myself with logbook data - I fought to halt my uncontrolled descent. Gradually I slowed to a stop. Then, kicking and clawing, I began to move upward. I shoved cables out of my way and rose straight up. As soon as I was free of the clutching mess, another thought assailed my overtaxed mind: what if I came up on the wrong side of the partition and couldn't see Starace's light? All I could see above me were bubbles.

It didn't happen that way. Suddenly, Starace's light burst into my face. I was never so happy to be blinded. I shot out of the hole without even brushing the serrated edges. As soon as I was clear of the wreck I hit the exhaust valve. I soared up another ten or fifteen feet before arresting my ascent, then fell back down to the hull next to Starace. He pointed to his timepiece. I checked my own gauges, and nodded. Not only was it late in the dive but I was low on air. I eagerly agreed that we had to leave at once.

We barely began the swim to the anchor line when Starace flashed his light at me from behind. When I glanced over my shoulder I saw that he was holding one fin in his hand. I went back to help. The strap was either broken or had come undone and could not be fixed on the spot, especially by one wearing quarter-inch three-fingered mitts. I tucked the fin under my left armpit while he fluttered along awkwardly with one leg.

By now I was very concerned about my shortage of air. I had to look out for myself. I overtook Starace and raced ahead for the anchor line. I kept looking back, making sure that he was in sight. When I reached the grapnel and started my ascent he was twenty feet behind. I kept my eyes on him as I climbed hand over hand. He signaled "okay."

I wanted to get out of deep water fast because that was where the most air was consumed. I ascended rapidly for the first fifty or sixty feet, then slowed down when I reached the one hundred feet. I could still see Starace below me. At a slower pace I moved up to thirty feet, my first decompression stop, then breathed easily and adjusted my buoyancy while holding onto the line with my right hand. Now, even if I emptied my doubles and pony bottle, I could reach the spare tank that hung down in the water on a line from the bow of the boat.

Starace soon caught up with me. He halted momentarily to check his gauges, then moved up to twenty feet. I had to decompress deeper and longer because I had attained a deeper depth. Suddenly I felt woozy. What I perceived as a blink of the eye was a momentary lapse of consciousness. I was roused from the swoon by the pain in my ears due to increased pressure. My thumb and finger still circled the rope, but I was sliding down the line because there was no strength in my grip.

My hand was paralyzed! The paralysis extended up my arm, along my right side, and down my leg. Deja vu.

I plummeted past fifty feet. I reached out with my left hand and grabbed the anchor line firmly while fighting waves of dizziness. Once I stabilized my position, I rolled over and looked up at Starace, who now hovered more than thirty feet above and spun dizzily like the knot on a lariat in the hands of a rodeo cowboy. He stared down at me and wondered what I was doing, but I had no way to let him know that I was in trouble and needed help. I closed my eyes and tried to shake off the disorientation.

The vertigo persisted. I was scared and desperately wanted to scream for the surface and the wrongly perceived safety of the boat. But logical thinking and instinctive reflex took command of the situation. I let myself fall down to sixty feet, reasoning that the added pressure would decrease the size of the bubbles causing the problem with my central nervous system. After my head cleared, I climbed up the line by kicking with my good foot and by working my fingers like caterpillar legs. I paused at fifty feet, worked my way up, paused at forty, reached thirty, and finally attained twenty feet. By this time Starace was hanging at ten feet, still nonplused at my behavior. He couldn't tell that half my body was paralyzed. He surfaced.

My tanks were nearly empty. Rather than drain them completely or use up the air in my pony, I lunged for the bouncing emergency bottle - not an easy task considering that half my body was paralyzed. Not only were surface conditions less than ideal, but the length of the tie-off rope had been miscalculated. It was all I could do to stretch the regulator down to the minimum depth of ten feet, and then the slight chop kept jerking the mouthpiece from my lips, often catching me unawares in the middle of a breath. I fought this situation for five or ten minutes, then gave it up as I was being worn out.

Life on the anchor line was easier. As feeling returned to my hand I was able to twitch my fingers. I forced - or willed - my fingers tighter with each succeeding twitch. After fifteen or twenty minutes of this I was able to make a fist - a weak one, but at least my condition was showing progress. I also began flexing my leg as the numbness there subsided. When I ran out of air in my mains, I switched to the pony, determined to stay down as long as possible until the cure was completely effected. Eventually I regained full use of my limbs: not a tingle remained.

According to my meter and Navy Table calculations, my decompression obligation was satisfied. But the medical abnormality obviated strict adherence to procedure. I decided to hang around a little longer and continue in-water treatment. I went back to the capering hang bottle, now not as concerned that I couldn't maintain a depth of ten feet. I hung more easily at eight. When the bottle was nearly empty I tentatively rose to the surface and drifted with the current to the ladder. I hung onto the lower rung for a few minutes, taking it easy.

Then I made an incredible discovery - I still had Starace's fin tucked under my arm! Since he was waiting for me at the top of the ladder, wondering why I stayed so long in the water, I passed it up to him. After climbing onto the boat I told him about my problem. I sat fully geared, waiting for tingles to reappear. None did. I was lucky.

It took all my strength and will power to forego the afternoon dive, the last of the trip. I remained on board, envious, while liftbags popped to the surface bringing up treasures from the deep. Burdewick finally sent up his porthole, and all together seven Promenade Deck windows were recovered, including mine.

The *Andrea Doria* was so huge, so fascinating, so incredibly alluring, that in three short dives it had attained a significance for me that superseded its status as a mere merit badge or rite of passage. I wanted to probe its hallowed hull and see more of its cavernous interior. I had counted coup by touching the wreck-diver's most sought after dream. But I wanted to return for the thrill of sheer exploration.

The trip to the *Andrea Doria* was undoubtedly the highlight of EDA's existence. Yet Roach lost interest in the wreck once he could say that he had been there. For me the wreck's appeal entered a new, more warranted dimension.

For the rest of the decade I tried to arrange charters to the *Andrea Doria*. Twice I came close to doing so. The problem was not enticing enough people to fill a boat, but finding a boat to take us. I almost had the *White Star* chartered, but upon reviewing his licenses Captain Ray Ettel found that he did not have Coast Guard approval to operate so far from his designated port. The boat docked in Barnegat Light, New Jersey. I *did* charter the *Amberjack II*, a fishing boat out of Sheepshead Bay, New York. But two weeks prior to departure Fred Ardolino shamefully admitted that he didn't really own the boat or call the shots as he had led me to believe - his father did, and his father was afraid of the liability of carrying divers. I was forced to return deposit money to a dozen disappointed divers.

PENETRATING THE DARK

For years the *Andrea Doria* resisted my strongest efforts to return. As a surrogate for what I later christened the Grand Dame of the Sea, I was introduced to another wreck that offered, if not the same mystique, an equivalent road to adventure at a much shallower depth. The USS *San Diego* was an armored cruiser that was sunk by a mine laid by the German U-boat *U-156* in July 1918. She was the only major U.S. warship lost in World War One. The wreck lay in 110 feet of water some eight miles south of Fire Island, New York.

The Navy had long abandoned the wreck to commercial salvors. In 1921, permission was granted to the Saliger Ship Salvage Corporation to salvage the wreck, although no records exist which establish that salvage was actually attempted. Saliger's claim was abandoned by the passage of time. In 1957, the Navy sold the wreck of the *San Diego* to Maxter Metals for the paltry sum of $1,221.00 "for scrapping only." This latter scheme was protested by the American Littoral Society, an environmental organization bent on maintaining the *San Diego's* hull structure as an artificial marine life habitat. Maxter Metals had so little money invested in the venture that it abandoned salvage operations voluntarily without compensation.

That the *San Diego's* extensive and convoluted interior lay entirely unexplored was patently obvious to me during my initial penetrations into the upside-down hull, when I found loose portholes and numerous other artifacts lying completely exposed only a few feet inside the yawning gun ports. The jungle trails of darkest Africa were better known than the collapsing compartments and intricately linked passageways of the *San Diego*, which gave the word "darkest" new meaning. This was the wreck in which I developed and honed my skills in shipwreck penetration - usually through successful trial, sometimes through near fatal error.

On the construction site of a skyscraper I noticed the rod setters pulling wire off a plastic reel in order to tie the rods together before the concrete was poured for the floor. This observation sparked a flash of inspiration: the wire could be replaced with string or line which could then be unspooled as I explored the interior of a wreck. I asked the rod setters where to purchase such a device, and they directed me to a commercial supply house not too far away. I went there during lunch, bought a reel and some polypropylene line, and converted my first wreck reel.

When the need arose, I laid lines along my route so I could find my way out even if the visibility was reduced to zero by suspended silt and sediment - like Hansel and Gretel leaving bread crumbs on the path to the witch's gingerbread house. Sometimes I left lights or continuously flashing strobes at entrance points in order to guide my retreat from a compartment or deck level.

But the method I utilized customarily I called "progressive penetration." This consisted of incursions made farther incrementally during a series of discontinuous dives, often weeks or months apart, enabling me to memorize details of complicated routes and far-in places, and to establish my position from known reference points despite disorientation. Thus I constructed a mental picture of the wreck from puzzle pieces collected and fitted together over an extended period of time, creating a map in my mind that could never be lost.

I suffered a few bad turns, to be sure. The worst was when I got lost in the after Berth Deck (a pun I got much mileage from) while diving with Carol Coleman, a school administrator. We entered through a square loading hatch less than three feet across. In those days, many of the studs or supports for partitions were still standing, so we had to wend our way through a maze of vertical beams in a somewhat circuitous passage. I blundered into a closet with Coleman on my fintips. As particulate matter stirred up and filled the enclosed space, and our exhaust bubbles dislodged sediment from overhead, I couldn't figure out where all the walls had suddenly come from. With visibility reduced to a couple of feet, I kept searching with my light and feeling with my hands for a way out of a box that had impossibly closed around us.

When I finally found a portal out of the prison I had no idea where we were or, worse, in which direction to head in order to exit the wreck. The cloud of silt seemed to follow us everywhere, making it impossible for me to obtain my bearings. I swam first one way, then another. Nothing looked familiar. I would admit that I was scared if I was, but I wasn't - I was terrified. I sensed that the end was near, and I prepared myself to accept death with resignation if not with courage. I didn't know how much air I had remaining, but neither did I waste the time to find out. Every breath counted. I made up my mind that when I felt the final drag and could suck no more air from my doubles, I would calmly switch to my pony bottle and continue to search for the exit. And when that air was exhausted I would not grab Coleman's regulator. I would not fight her for her air or ask her to share, but would expire quietly, still looking for a way out. That is how sure I was that I was going to die.

I turned to Coleman and held out my hands to indicate my confusion.

Always a gutsy gal, she shouted quite clearly through her regulator, "Stop fucking around!"

Her confidence impressed me. I knew that it was up to me to find an escape route, but I was afraid to move for fear of penetrating farther into the wreck and into unknown territory from which there was no hope of egress. The only way I knew out of the Berth Deck other than the way we came in was down the skylight opening that led to the Gun Deck - but everything below knee height was obscured by a thick layer of silt. In desperation, I took a bearing off my compass and followed the needle due west. This action was based upon my knowledge that the wreck lay approximately north-south.

By maintaining a straight course we soon emerged from murkiness into crystal clear water. I was emboldened by good visibility. I didn't recognize the area, but when I spotted the outer bulkhead my heart throbbed with relief. Now we only had to follow the steel barrier till we tumbled out through the square hatch. My best guess was that we were forward of the opening and had to turn left to find it. I soon discovered that I was wrong. We were farther aft than I had ever been before. I could have turned back immediately, since the square hatch could not be more than fifty feet away - a minute or two in travel time - but then I noticed a glow of light ahead. Structural failure had caused the hull plates to separate, and the slender gap was barely wide enough to slip through. A moment later we were outside the wreck.

I dashed madly for the anchor line a hundred feet away, waving to Coleman to follow but not waiting for acknowledgment. I was nearly out of air and had no time to engage in hand signal communication. I squeaked out enough air to complete my decompression without having to breathe from Coleman's ample supply - she was "good" on air and always surfaced with her tanks half full. She climbed up the ladder right behind me. I slid across the bench so there was room for her to sit. Now she was angry, and wanted to know why I was fooling around inside the wreck.

I said, "I wasn't fooling around. We were lost."

She removed her mask to reveal a perfectly blank expression. "Oh. Were we lost?"

I regained enough composure to say, "Ignorance is bliss."

Years later I had another major scare in the *San Diego*. I took with me Kathy Warehouse, a draftsperson, to act as my safety while I slithered through a mud-filled slot which I believed, from examining the ship's plans, was connected to a compartment I normally accessed from a more roundabout route. The main problem was that the decks in the area were collapsed and rusted through, making it deceptively easy to change deck levels unknowingly, and to wind up somewhere that was not part of the plan. I stationed her at the opening with instructions to stay put and point her light after me as a beacon for my return. Since I had to crawl under a fallen partition with my tanks and belly scraping, I couldn't avoid churning the passageway into thick sediment soup. I squirmed between beams and over metal debris to the other side of the wreck, where I thought I recognized points of reference made from previous exploration, although the appearance was quite different because my viewing angle was reversed.

I wasn't confident enough to proceed, so I went back to the silt-filled crawl way and looked for Warehouse's light. I couldn't find it. I groped forward tentatively, leading the way with my own light stretched out in front of me and waving the beam slowly back and forth. There was no answering gleam or glow. I lay in utter darkness for perhaps a minute or two while improbable scenarios that explained her disappearance passed frightfully through my mind. Afraid that an attempt to traverse the muddy obstacle course would end in my becoming irretrievably lost in an unknown sector of the wreck, I opted to back through the engulfing silt into territory that demonstrated a semblance of familiarity and - more important - a possible connection to the open spaces in the stern.

I regained clear water, adjusted my buoyancy so as not to drag dangling gear on the bottom, and - my heart thrumming hard against the wall of my chest - finger-walked along the outer bulkhead in what I hoped was the right way out. Every finger-step brought increased recognition. Soon I felt fully confident of my whereabouts. Then it was merely a matter of traveling a well-worn path around formerly marked obstructions to a transverse corridor that led me out of harm's reach. I was back in home territory.

But when I reached my initial take-off point deep inside the wreck, Warehouse was still nowhere in sight. I thought she must have gone after me into that swirling black abyss - just as I would have done in her place. In again I plunged. Once more I lay in ebony solitude, playing my light back and forth, only this time peering in instead of out. I crept ahead a foot at a time until I gained the partial clarity of the room that I had vacated just minutes before. There was no sign of her. For a second time I finger-walked the long circuitous route, until I circumnavigated the interior completely a second time and entered the large open space where the adventure had started so expectantly. Still no Warehouse.

Just as I feared the worst about her possible hideous fate, she happened along holding a lobster in her hand, completely oblivious to the endangerment she had caused. Afterward, she freely admitted abandoning her post and looking for lobsters during my absence. To my stern words of recrimination about dependability and her responsibilities as a buddy, she responded with indifference. It was not in her nature to care about other human beings unless their existence affected her in some way. I never dived with Warehouse again.

The Collapse of EDA

Meanwhile, EDA became pedestrian and autocratic. No longer a small group of elite, Roach expanded the organization into a money-making enterprise that catered to mediocrity. More than a hundred members swelled the rolls in 1975. The schedule was padded with inshore dives to mollify new recruits who joined the ranks to gain experience in depth and decompression, but who found themselves uninvited on the wrecks that really counted, and who were relegated to the role of slot-fillers on undersubscribed charters whose purpose was to stuff the club's coffers with cash. This bait-and-switch routine was unfair to the majority, who unknowingly supported Roach's "private" charters available only to those in the inner clique. It created dissension among the ranks.

I was very much a part of the inner clique. In fact, I was Roach's sole confidante, on whom he relied for advice on all club matters. I was also responsible for evolving club policy, and I suggested most of the changes that were ultimately incorporated. Being uncredited advisor and club vice president without title didn't bother me in the least because I had no interest in exercising control over my fellow divers, nor did I have any desire to lead the members by force of will. I just wanted to dive, set a good example, and create an environment for those with a similar bent.

Roach, on the other hand, wanted to be supreme commander - and he was quite willing to acknowledge his base delusions of grandeur. He jokingly called himself the "king" so often that the long-time members - his peers - gave him an ornamental crown. The newcomers thought this was a gesture in jest, but those of us who knew Roach better saw through his thin comedic facade. He truly believed that he was the best and that he deserved to rule others who were yet less skilled. This despotic drive eventually undermined his leadership and proved to be his downfall.

Gone were the days of cheerful camaraderie, the days of harmless high jinks: such as salting a wreck with kitchen china, or chipping in to buy a flea collar for Don Nitsch when he resumed diving after a serious bout of Rocky Mountain spotted fever. Now there was a struggle to curry favor with the king in order to get on special trips.

EDA had become a business.

The turning point in EDA's rise to respectability was the Gafney incident. In addition to belonging to EDA, Paul Gafney was a member of Main Line Divers, and knew quite well the charter rates of the various boats. Dive clubs ordinarily held open forum discussions among the general membership, so that all members were kept apprised of the club's financial position. The treasurer's monthly reports detailed income from dues, raffles, and other fun raisers, and balanced the budget by offsetting expenses such as charter fees, meeting room rental, newsletter printing costs, and so on. The condition of the treasury was open knowledge. Since dive clubs were non-profit organizations, the cost of a boat charter was divided equally among the participating divers, leaving a net gain of zero. In the event that a boat was not filled to capacity, the difference was taken out of the club treasury. The primary function of a dive club was to schedule dives and assume responsibility for financial losses.

Gafney made some simple calculations and discovered that in EDA a great disparity existed between the cost of a charter and total amount collected from participants. If divers were being overcharged, he wondered, where was all the extra money going? EDA held no meetings, published no newsletter, and supported no events such as parties, annual dinners, or charity fund-raisers.

It was common practice then and is more so today for the dive master to go for free as a way of reimbursement for leading a trip. The deficit thus produced was compensated by increasing slightly the charge per person, or by adding the dive master as an extra person beyond the prescribed optimal number.

Roach didn't pay for his diving, but he spent a great deal of time running EDA by himself, so no one objected to subsidizing his habit. As the organization grew, the long distance charges on his phone bill became enormous, so he padded the head fee to make up for his expenses. Still no one minded. But somewhere along the line he got greedy. From excessive increases in the cost per person he charged, it became apparent that Roach had been suborned by the profit motive. It was this skimming and scheming that Gafney stumbled onto and, by the enmity he received from Roach's recriminations, brought harshly to the attention of the members.

Roach's creative price-fixing formula could not survive a strict accounting. Nor could his high-strung ego endure an assault upon his sovereignty. He was president, treasurer, secretary, and sergeant at arms, all rolled up into one. He was beholden to no one. That Gafney should have the temerity to challenge his authority was more affront than he could handle.

I recognized the equity of Gafney's position, and saw him as another underdog who needed defending from the pack, so I tried to mediate an understanding between both sides, completely without success. Since I belonged to several clubs and was aware of how they operated, I explained to Roach why Gafney's question was one of innocent curiosity. I explained how the issue of finances was treated with openness and honesty in other clubs. And I explained how retribution was not a procedure to adopt.

Roach accepted none of it. He no longer viewed EDA as a democracy or as an association of divers with equal rights, but as his own private machine with himself in sole command. He suffered no insubordination from inferiors.

Rather than answer to Gafney's unintentional allegation of impropriety, Roach returned his escrow money and banned him from future dives. Thus Gafney earned the distinction of being the first member to be kicked out of EDA. This action, executed without committee sanction or a vote from the membership, made all members demonstrably aware of their own vulnerability should they incur Roach's wrath. A sense of unease disturbed the ranks and "the wrath of Roach" became a buzz word.

I championed Gafney's cause, counseled Roach firmly on the injustice of his deed, and pleaded hard to have Gafney reinstated. Roach remained arrogant and obdurate. I persisted alone in my protestations. No one wanted to side with me for fear of being punished with a similar fate. To me, however, ostracism was not as important as principle. It didn't even come close. I continued to exert pressure on Roach until he finally reached the breaking point. I knew that I was pushing him too hard, and knew as well the inevitable result, but I wouldn't back down from my position.

I too got the boot (or the fin) and had my escrow returned.

These incidents occurred in 1976: the nadir of EDA's troubled existence and the beginning of the end of its influence. Roach was now viewed by most as a tyrant with a badly tarnished image. Not only did he earn the enmity of his most respected peers, he lost prestige from those he considered beneath his glance. If I, Roach's most esteemed confederate and number two person in EDA's totem, could be kicked out of the "club" on presidential whim, what prospect could the lower orders expect?

The schism that my departure created could not be repaired. Although I had some competitors in EDA, I had no enemies; and most everyone but the latest arrivals, whom I had not yet gotten to know, was my friend. I thrived on friendship. EDA limped on for a while in weak imitation of its former glory.

Grand Artifact Display

Apart from the divisive issues that threatened to tear the fabric of EDA into shreds, there occurred an event that brought wreck-divers

270 - The *Lusitania* Controversies

together from clubs over the entire four-state area (New York, New Jersey, Pennsylvania, and Delaware). This was a weekend artifact exhibit held on an old ferry boat that had been converted into a restaurant and was permanently docked in Brielle. The dates of the extravaganza were March 20-21, 1976. I don't know who was responsible for the logistics, but a great deal of time and forethought went into contacting divers who had unique artifacts in their collections, allocating space for booths, covering tables for displays, making signboards, decorating, and handling the artifacts as they arrived. It was a monumental task accomplished strictly by volunteers.

Practically every major shipwreck was represented and featured in its own booth. Lesser dived wrecks received recognition by dint of descriptive plaques. From as far north as Rhode Island there were items recovered from the twin victims of World War Two, the collier *Black Point* and the *U-853* which torpedoed her. From as far south as Maryland came the *Washingtonian's* fog horn, which Dave Ford tooted with compressed air piped into the resonating chamber from a scuba tank. (Dave Ford was a Dupont employee who was no relation to the Gary Ford who died on the *Cherokee*.) The Aquarians brought their complete collection of artifacts from the Revenue Cutter *Mohawk*, including the bell and a brass gauge panel, both imprinted with the ship's name. Standing on its own was the helm from the *Ioannis P. Goulandris*. From the ancient wreck of the *Western World*, lost in 1853, was Paul Hepler's huge capstan.

From wrecks of all ages and descriptions there were deadeyes, portholes, telegraphs, lanterns, cage lights, valve wheels and assemblies, bottles, glassware, china plates, ceramic mugs, speaking tubes, capstan covers, windows, compasses, steam gauges, telescopes, sextants, blocks, and every imaginable ship's part that had managed to survive the terrible onslaught of nature's corrosive bath. Every object was wonderfully preserved, polished, and presented by the people who cared about them the most: wreck-divers.

One item that brought particular attention - and open guffaws of laughter - was the bell from the *Balaena*. Hoffman had made it into the centerpiece of a lamp. Someone - and it was never revealed who - removed the cardboard name plaque on which was written "recovered by George Hoffman" and replaced it with one that stated "found by Joel Entler." Although word of the verification of truth spread like wildfire among the guests and participants, Hoffman never noticed the switch, and the truth persisted.

The gala carried on late into Saturday night with much laughing, drinking, and carousing - the latter mostly in private, but only mostly. It was the largest collection of local maritime artifacts ever assembled - before or since - and it clearly demonstrated the love that wreck-divers have for their hobby, and the effort they were willing to put into preserving the history that they worked so hard to recover.

Plans for subsequent annual exhibits never materialized because the ferry boat burned, sank, and was scrapped.

Striking Out on My Own

The year also found me moving in other directions. Within the protective enclave of EDA, deep dives and long decompressions were commonplace. But certain people in other clubs disdained such practices. These were people who felt that their status within the club was threatened by the accomplishments of others. They sought to remedy the situation by denigrating diving activities that were considered beyond the norm (read: beyond their ability).

On club trips there was no way that I could dive deeper than anyone else, but I always stayed down longer. I was the first one in the water and the last one out. While decompressing, I waved to people as they descended the line, then waved to them again on their way to the surface. These decompressions were no longer than those done on EDA trips. The difference was the setting. People who were new to the sport or who had reached their limitations looked upon me as a madman: a reputation that I carried for many years.

Not completely satisfied with the destinations offered by the clubs, I chartered boats to take me to the wrecks I wanted to explore. I found enough disgruntled people, who wanted to dive deep and long and without political affiliation, to accompany me. I also ran my first wreck-diving trip out of the country, to Bermuda, where shipwrecks had been piling up on the shallow water reefs since 1607, when the first settlers landed accidentally on the arid, hostile shore. I ran many more trips to Bermuda thereafter.

A New Wreck-Diving Influence

The most important development of the year, for me personally as well as for wreck-diving in general, was the opening of The Dive Shop of New Jersey, which was owned and operated by Norman Lichtman. He had previously been in partnership with Fred Neuman in Aquatic Recreational Enterprises. After Lichtman and Neuman closed A.R.E. due to difficulties in the partnership, Lichtman struck out on his own. He had a vision that a dive shop should do more than train people to dive and supply them with gear. It should provide a mechanism by which people could increase their proficiency beyond the basic skills taught in introductory classes.

The certifying agencies offered specialty certifications for such underwater activities as night diving, ice diving, river diving, and so on, but normally the curriculum consisted of little more than a three-hour lecture and a single baptismal dive, after which the graduate was permitted to wear an identifying patch that denoted successful completion of the course. Ordinarily these bare minimum courses appealed to people who collected patches like a Cub Scout accumulates merit badges, and who plastered them on their sweatshirts or jackets as if they were combat ribbons. Ordinarily, too, these overachieving divers rarely pursued the special activity once they earned the patch, since their goal was not to hone their skills but to display their imagined prowess. I called these divers "patch people." The disparagement is intended, since the number of patches a person wears is not a measure of his skill.

No one at the time taught wreck-diving, which seems odd when I look back on it because that was almost the only kind of recreational diving being conducted off the Jersey coast. Basic open-water scuba gave a person a license to get his tanks filled, but didn't truly prepare him for the rigors of the ocean or the hazards encountered on sunken shipwrecks. People learned the hard way, as I did, by "on the job" training. Lichtman sought to change all that.

He began his program by soliciting my aid. He introduced an advanced course that was geared specifically toward shipwrecks. For the penetration segment he had me install a guideline inside the *Stolt Dagali*, then give the students a prep talk on the boat and escort them through the darkened corridors. Sometimes he took them through himself. Once they learned the ropes in this manner they gained a confidence and level of skill that had taken me years to attain. It was faster to be shown than to reinvent the wheel. I never wanted to be an instructor, but passing on my experience in the field (or in the water) I found satisfying. I liked making the opportunity available for people to discover for themselves what wreck-diving was all about. It was a way of sharing life's experiences.

The next step was to teach decompression diving. As with wreck penetration, Lichtman tutored the divers in class on theory and on how to make and deploy line reels, then turned them over to me or to one of his teaching assistants for the actual in-water session, or came out on the boat himself. I didn't hold anyone's hand. I simply gave them a last minute review of procedure, inspected their equipment, and let them handle the rest. In order to circumvent the liability threat, and to ensure that no one got hurt, the students trained by doing only a simulated decompression dive - one in which they had not exceeded the no-decompression limit. That way, if someone snagged his line on the way to the surface, he could ascend without fear of getting bent. This kind of preparation and familiarization would have saved me some anxious moments on that long-ago dive on the *Bass*.

But this was only the beginning. Lichtman also wanted to run dives through the shop the way the clubs did, in order to keep people diving after he certified them. Odd as it may seem, the majority of certified divers failed to pursue the activity after completion of the course. This occurred partly because some people found that diving didn't really appeal to them, but mostly it was because no opportunities to dive were made available to them. Many instructors, after collecting the course fee and effecting the sale of the basic scuba outfit, promptly forgot about their graduates and went on to a new batch of students. My own training experience was an example.

Lichtman's philosophy was just the opposite. He believed that an instructor had a further obligation to his students. I won't ignore the profit motive of such an attitude: the more a person dived, the more equipment he rented or purchased, which augmented business for the shop. But in this case everyone benefited. Also, contrary to the contemporary mindset of the certifying agencies, Lichtman was a staunch supporter of the pony bottle as a diver's best and most reliable buddy, a concept that was heretical to the old-school establishment but which had long been commonly acknowledged by those who did more "real" diving than teaching or talking about it.

Most people take up diving primarily so they can see for themselves the vast array of marine life that inhabit tropical coral reefs. For them, Lichtman provided exotic vacation packages to the Bahamas and other luxury resorts. These people are largely once-a-year divers. But wreck-divers were different. They wanted to dive locally on the weekends, and with regularity. It was to satisfy this latter group that Lichtman asked me and Bart Malone to produce a dive schedule for the shop.

Bart Malone was a carpenter who lived near the shop in New Jersey. Together we made all the arrangements by calling the various boat captains, choosing dates, and selecting destinations that would appeal to divers of varying skill levels and with different interests: from those just out of class to those with years of experience. We picked our favorite wrecks: from shallow to deep, from easy to challenging, for digging and for penetrating. One or both of us went on the trips as dive master. We were not salaried employees of the shop and received no compensation for our labor. But neither did we have to pay for the trip.

At first, Malone and I had to actively solicit divers to fill the charters. We spent a fair amount of time on the phone. But as word got around and as Lichtman increased the number of training courses he taught, the charter initiative became self-sustaining, until it reached the point where trips were fully subscribed months prior to the date of departure, and people pleaded to be put on the stand-by list, to be called in case of a cancellation. The cost per person was competitive with what the dive clubs charged, with the additional advantage that there were no dues to pay. (As a reminder, in those days an all-day offshore trip cost $15 per person; inshore cost $12.)

The charter schedule was not a profit-making proposition. In the early years it ran at a loss. But the revenues it brought in to the shop in the way rentals and equipment purchases more than made up for any deficits.

As I did with EDA, I introduced the escrow system. By a defect in human psychology, there are people who believe that when they fail to fulfill an obligation, someone else should have to pay for it. By such thwarted logic there are those who don't want to pay for a missed dive, regardless of whether they overslept, got lost on the road, had transportation problems, had a hangover, or just didn't feel like getting up that morning. These people wanted the charterer to pay for their spot. The escrow system solved that problem.

Within a couple of years The Dive Shop of New Jersey was teaching eight nightly courses per week. Lichtmen had to train other instructors to handle the load. Gene Peterson, one of Lichtman's instructors and now a dive shop owner himself, taught three classes a week, some with as many as forty students. A typical check-out day at the quarry might have thirty students attending. Lichtman rented a bus to transport all the students together, and provided refreshments and a picnic lunch for the outing. He certified nearly four hundred new divers per year. Equipment sales and rentals soared.

We enlarged the schedule to meet the demand. During the prime summer months there was a trip scheduled for every Saturday and Sunday, and some weekends had three trips scheduled. Days were either doubled up with deep and shallow destinations, or a night dive was fitted in between. In addition, during the height of summer there were trips scheduled on Wednesdays, for vacationers and for those with flexible work schedules.

INSTILLING THE CODE

By catching people early in their diving careers, usually straight out of class, we had the perfect opportunity not only to introduce them to the wonders and hazards of wreck-diving, but to demonstrate the codes and ethics it was necessary to instill in light of society's confusing values. One representative incident that comes to mind occurred when a diver brought up a porthole that another diver had just removed and tagged but had to leave behind because he ran low on air. He came up for a second tank. Another diver who happened along soon after, in his exuberance over finding a loose porthole, failed to notice the first diver's lanyard, and sent the artifact up on a liftbag. This created a quarrel on the boat that I had to adjudicate before a fight broke out. I listened to both sides of the argument, examined the porthole, and pointed out the lanyard to the second diver. He acknowledged its presence and admitted that he hadn't noticed it underwater, but he still refused to back down from his position of ownership since he had, after all, sent the porthole to the surface.

After patiently explaining the diver's code of priority, and finding myself unable to appeal to his higher principles in the excitement of the moment, I offered to *give* him a porthole that I had recovered previously from the same wreck, in order to mollify him. That shocked him into accepting the seriousness of the matter and understanding how strongly I felt about adhering to the code. I particularly did not want to set a bad example of fairness or create an unjust precedent that might affect future deliberations, or to send an unclear message to the diving community in general. After tendering my proposition, it took him only a moment's reflection to accept the proper verdict. He shook hands with the other diver, apologized for not noticing his lanyard, and agreed that he should have the porthole. He also refused to accept my porthole as a replacement.

Lichtman, Malone, and I embraced other trusts as well, such as easing people through their formative early dives and the development of skills, when they lacked confidence in their performance and were vulnerable to feelings of insecurity. One bad experience or the lack of sensitivity from an authority figure could easily turn off a person so he might quit the sport or never achieve his full potential.

For example, one beginner climbed on the boat in complete disgust because he'd been unable to get down the anchor line past decompressing divers. The current was running strong enough that he couldn't let go of the line without being swept away. "They weren't doing anything," he complained. "They were just hanging there blocking the way." My impulse was to shout, "You stupid jerk, don't you know that they're decompressing, and that they are at significant risk?" But I didn't. Instead, I controlled my temper, elucidated the importance of staged ascents, and suggested that instead of putting away his gear, he go back in the water after they surfaced. He did, and had no further trouble. His name was Angelo Patane, and he went on to become a highly skilled diver. It's almost frightening for me to think how an attitude of intolerance or insensitivity has the awful power to crush self-esteem at a crucial moment, so that recovery becomes impossible.

DIVE MASTER DUTIES

As dive masters, most people saw us counting heads and collecting money, but our primary responsibility on the boat was safety, a commodity not so noticeable unless one happened to be in trouble. Rescues were not uncommon. I dragged my share of unfortunates back to the boat, or swam ropes out to them so they could be hauled in ignominiously like limp fish at the end of a line. Although many people have thanked me graciously throughout the years for saving their lives, I doubt that many of them would have died despite the lack of immediate assistance. They simply would have had to wait longer for help to arrive - may perhaps have floated off into the sunset and had some anxious or terrifying moments they could have done without. I suspect that in the panic of near drowning one may perceive death to linger only a breath or so away. "Rescue" is more often a case of "relieved from stress" than "saved from death." It's all a matter of perspective.

Most problems originated from fast current. In those days, no one had yet thought of securing a rope to the anchor line so a diver could pull himself hand over hand to the bow instead of kicking with every ounce of strength he could muster till he was breathing like a race horse that had just lost the Preakness. Add to this the fact that the average person was somewhat out of shape, and might be wearing a rental wetsuit that fit too tight, and you have a recipe for rescue.

On the larger boats that carried twenty people or more, Malone and I found that the two of us could easily become task loaded, especially when the seas were rough or the current was strong. When one of us dived, the other stayed fully dressed on watch on the bow. But if that person had to conduct a rescue, there was no one left on board to go after the next person in trouble. Lichtman came up with the idea of having additional safety divers, or "safeties," as we called them.

Safeties were paying customers of exceptional standing who volunteered their services. It was no bargain to sit in the sun fully geared in a drysuit and tanks on a hot August afternoon - and to pay for it to boot! But divers like Lynn DelCorio, Dave Poponi, and Harley Sager did just that, and earned my everlasting respect. They and others like them went on to become dive masters or instructors, and carried on the tradition with The Dive Shop of New Jersey when I passed on the torch after eleven years of service, in 1986.

DIVING BECOMES FUN AGAIN

Dive Shop trips were great fun, full of camaraderie, and totally lacking in the vicious competition that characterized EDA. More often than not, divers helped one another to recover artifacts, or engaged in team efforts to work on projects that one person couldn't handle alone. People treated each other with respect despite varying degrees of skill. I have many fond memories from those days.

One time the "gang" thought to play a joke on me by slipping a doughnut into the bootee of my drysuit between dives. Most everyone was in on the gag. The tricksters (Gene Peterson and Lynn DelCorio) snickered and waited for the moment of delight that would come when I made the mushy discovery. Then they shrank in horror as Mike de

Camp reached for the drysuit and began to pull it on. The reader must understand that de Camp was revered by the masses as a guru, an idol, an untouchable, with almost godlike status. One did not play pranks on the deities.

I knew otherwise. He was an ordinary guy with extraordinary drive and skill, sometimes given to clumsiness. Once he threw his doubles on the gunwale with such vigor that they tumbled overboard, and someone had to retrieve them from the bottom. Another time he dropped his camera rig in the water. Hoffman tossed a buoy on the site, and Roach and I went down to 130 feet to look for the very expensive equipment. Roach remained at the drop weight holding the decompression reel while I swam around in ever-widening circles. Suddenly, Roach pointed vehemently over my shoulder, but when I turned around I didn't see anything. I shrugged, and a moment later I found the camera and strobe on the sand, which I assumed he had been indicating. Already a lobster had taken up residence under the strobe arm, so I scooped it up with the camera rig and put everything into my mesh bag. De Camp got dinner as well as his photo equipment. Not until we climbed on the boat was Roach able to communicate that a large shark had passed right behind me, once again proving the old adage that ignorance is bliss.

De Camp's drysuit was the same brand and color as mine, which the pranksters on the boat failed to notice. They slunk away like whipped curs in anticipation of an outburst of temper. But because de Camp wore thick foot protection for warmth, he failed to notice the squashed pastry at the time, so they were saved from retribution. Everyone involved was sworn to secrecy. Not until months afterward was the sordid story revealed to me by Gene Peterson, who was by then one of the managers of The Dive Shop of New Jersey.

SERIOUS SNAFU

Lest one be lulled into supposing that all shop trips were fun and frolic without misadventure, I offer the following incident. One spring dive in 1977 was the first of the season for many. Dive gear packed in the basement for the winter first saw the light of day on the boat. Dave Bullock unfolded his drysuit to learn that the inflation valve was sticky. I gave him a can of silicone spray to loosen the metal parts. He worked the valve free, donned his drysuit, and went in the water with two buddies. Five minutes later he appeared behind the boat, climbed up the ladder, which on the *Capt. Cramer* (out of Stone Harbor, New Jersey) often required help. I lent him a hand until he got comfortable on the back bench. Then he slipped out of his tank harness and sat quietly catching his breath.

Before I had time to ask why he cut his dive short, he fell forward into my arms and went into convulsions. I eased him down on his side on the bench, pushed his protruding tongue into his mouth with my thumb, and held it out of the way so he could breathe. His body didn't go rigid as it would if he were having a seizure, but stayed limp. He convulsed for a couple of minutes without vomiting. He did not regain consciousness.

I shouted for Don Cramer, the captain for whom the boat was named, to call the Coast Guard on the radio. He demurred. Bart Malone leaped into the crisis with his usual determination. He stood by Cramer's side in the wheelhouse and forced him to make the call that he was unwilling to make on his own. Someone told Cramer that Bullock had had a heart attack, and this is what Cramer related to the Coast Guard.

For a while Cramer relayed questions and answers through Malone between the Coast Guard and me, but too much information was getting lost in retelling, and the Coast Guard was balking about the necessity of sending a helicopter for emergency evacuation. The Coast Guard wanted us to recall our divers and bring in the patient ourselves. I was livid.

Where the Coast Guard picked up the notion that we could recall divers like whistling for a dog in the woods, I have no idea. The *Northern Pacific* was 140 feet deep, and divers would require decompression after only ten minutes on the bottom. We had to convince the Coast Guard that this was a matter of life and death, and that Bullock had to be flown to a recompression chamber at once. Every minute counted. But we couldn't convince the Coast Guard through an intermediary who didn't appreciate the gravity of the situation, and who was afraid to overrule a Coast Guard recommendation because his license to operate a commercial vessel was issued by the very same Coast Guard. Cramer was not one to take the bull by the horns.

I had to leave Bullock in other hands. Fortunately, a nearby customer took over the task, and I wish I could remember his name because he deserves credit for doing yeoman's service by keeping his thumb on Bullock's tongue until he left the deck in a stretcher. I took the microphone from Cramer and told the Coast Guard operator in no uncertain terms that we needed a helicopter *now*, and that I would take personal responsibility for making the demand. That got the operator's attention, and a helicopter was scrambled.

Malone and I directed the clearing of the deck of all loose gear so the down wash from the rotor blades wouldn't blow anything overboard. Although we were anchored some forty miles at sea, the helicopter arrived in comparatively short order. We cautioned everyone not to touch the descending stretcher until it grounded on the rail, lest the static charge cause electrocution. Bullock was still unconscious when we strapped him in the stretcher. The pickup went off without a hitch. But that was by no means the end of the story.

I gave the Coast Guard implicit instructions that Bullock needed treatment in a recompression chamber, the closest one being at the University of Pennsylvania in Philadelphia, and directed that he be flown there nonstop. Instead of heeding my advice, they flew him only as far as a beach community, where a waiting ambulance drove him to a local hospital with no chamber facilities and where no hyperbaric doctor was in attendance. This was intolerable.

Granted that we couldn't swear that Bullock's medical condition was pressure related, but that was an assumption that couldn't be ignored under the circumstances. Malone headed for the hospital as soon as we hit the dock. The situation he encountered was worse than expected - was, in fact, flagrantly criminal. The attending physician knew nothing at all about diving related injuries and was treating Bullock as a non-diving patient in coma of unknown etiology and whose condition, in his opinion, was stable.

It would be gross understatement to write that Malone was insistent or that he argued persuasively. He was as tempestuous as a winter gale. He fought bitterly to have Bullock moved to the University of Pennsylvania. He even offered to drive him there himself. Malone was adamant: he refused to leave the hospital or to let the doctor alone until arrangements were made for Bullock's transportation to a recompression facility where he could receive the proper treatment. Not until late in the evening did the doctor concede to his demands. An ambulance was provided, and the chamber facility in Philadelphia was notified to prepare to receive the still unconscious patient.

Bullock regained consciousness after the long-delayed hyperbaric therapy, but he was paralyzed. He didn't walk for months, and his recovery was a long, slow process. After six months he was still on crutches; a year after his accident he was able to get around with the use of a cane. He still limps today.

Bullock's dive buddies said that they lost sight of him on the anchor line, implying that he never reached the bottom. I never spoke with Bullock afterward, so I don't know what story he told about why he scrubbed the dive, but I have always been haunted by the possibility that his inflation valve jammed open and he was blown to the surface out of control. It is now commonly understood that too fast an ascent, even after a no-decompression dive, can cause the bends.

End of an Era

While The Dive Shop of New Jersey was growing, the Eastern Divers Association was falling apart at the seams. Fewer deep dives were scheduled, and Roach failed to participate on many of the "common" charters, assigning someone to oversee the trip in his stead. An aura of distrust glowed in the ranks, fueled largely by my unfair dismissal.

Roach undoubtedly was incensed when I chartered the *Sea Lion*, not just to dive but to look for new wrecks. My search for the U.S. submarine *S-5* failed in its primary objective, but the three-day quest resulted in the discovery of both the bow section and the bridge of the badly shattered destroyer *Jacob Jones*.

When 1976 came to a close, all but the eulogy had been delivered over the barely surviving corps of EDA. Roach found it difficult to fill the boats despite a severely reduced schedule. He scrubbed charters that were undersubscribed, sometimes without telling the people who were signed up for the trip. They learned the disappointing truth when they showed up at the dock and were turned away from the boat, which was chartered to another group. In 1977 he canceled all remaining charters and disbanded forever on the Eastern Divers Association.

An era had come to an end.

The organization that Roach expanded with such promise terminated in disgrace. He hung up his tanks and never dived again. Gafney and I were the lucky ones because we got our escrow back. No one else did.

Roach never understood that the most effective leaders spread influence through example. Those who try to rule by force are doomed to failure.

My portrayal of Roach's character is unfairly one-sided. It is primarily based upon our relationship and the differences in our personalities. I fully understand that I have a hang-up about being manipulated or controlled, and that others are not necessarily bothered by such abuse or exploitation. Roach was not an evil person. He was a doting husband, a loving father, a dedicated worker, and fair in his dealings with people who stayed in their place without complaint. In a sense I failed him by not being suited psychologically to accept his friendship under the terms with which it was offered.

Another Artifact Exhibition

That autumn, the Philadelphia Civic Center sponsored an artifact show that rivaled the one held in Brielle. Fewer artifacts were exhibited, but those that were displayed were shown off in spacious settings accompanied by professionally printed placards and enlarged photographs of ships and wrecks. Because the exhibit lasted for several months, untold thousands of people, divers and non-divers alike, got to see and appreciate relics recovered by wreck-divers from the area. I contributed much from my growing collection, and worked with the director of the exhibit in tracking down specific artifacts that I thought would make a valuable contribution to the display. I also suggested that he contact Tom Roach. He did, but Roach didn't want to have anything to do with the exhibit.

Transitional Phase

Although I was not personally affected by EDA's passage, the dissolution created a partial vacuum in the deep diving realm. Those who relied solely on Roach for scheduling trips were left in limbo. The Dive Shop of New Jersey absorbed some of the Garden State members, but for the New York contingent the south Jersey trips were a long way to go. I added more *Sea Lion* charters to the schedule in order to compensate for the fractionation created by the spread of geography, but this was only a bandage on the injury that Roach had inflicted. Understandably, Lichtman didn't want the shop schedule flooded with extremely deep dives.

Quite a few people wanted me to pick up the pieces of EDA and glue them back together again, under a new name if necessary. I gladly accepted the vote of confidence, but I was missing one important ingredient: the muster role. Odd as it may seem, I had no idea where most of the members lived, even those I counted among my closest friends. EDA consisted of people from all walks of life, who were drawn together by their uncommon interest and not by their proximity to one another, as is usual in clubs. The people I met on the boat I met *only* on the boat. Further social contact was precluded by residential distance and by the lack of ordinary club functions.

At that time I also dived with two Delaware clubs (Delaware Underwater Swim Club, and Dupont Employees Skin and Scuba Diving Club) and with the Atlantis Rangers in the Washington, DC area. Their loci were completely out of my bailiwick. I didn't belong to the clubs so I didn't have member's privileges, but I maintained contact with their dive masters by telephone and signed up for undersubscribed trips destined for wrecks I wanted to explore. Although I met some of these people only once a year, or perhaps every couple of years, our friendship was such that, whenever I met an acquaintance on the boat, I felt as if we had just been together the previous week: we simply picked up our association where we left it off, as if no great time had passed. It was pretty much the same with faraway EDA members. Our focal point was the dive boat and shipwreck destinations, but we were removed from each other's personal lives.

There was no chance that Roach would give the membership list to me, so Danny Bressette tried to get it from him. Despite his close ties with Roach his efforts were in vain. Roach didn't want EDA resurrected. If he couldn't be king, he wanted the kingdom destroyed. Together, Bressette and I managed to reconstruct part of the muster role by calling people who knew other people who knew other people, but recent additions and those who dived infrequently were hard to track down.

With the few hard-core divers we were able to salvage, we ran trips to the Mud Hole, primarily aboard the *Kiwi* out of Shark River Inlet, New Jersey. The *Kiwi* was owned by Floyd van Name, who also owned a dive shop in New York City that he later expanded to a commercial diving operation. The *Kiwi* was operated by fun-loving, beer-drinking Captain Mick Trzaska, who worked full-time for the New Jersey Turnpike Commission. With true gallows humor, Trzaska posted a sign on a pile next to the boat, which read: "Scuba - Some Come Up Barely Alive." I also chartered the *Kiwi* for The Dive Shop of New Jersey, for shallower dives.

On one *Kiwi* trip I was the last diver on the wreck when the grapnel broke free. I tied off my line and unreeled to the first decompression stop. Later, I surfaced to find that the boat was nowhere in sight. The horizon appeared blue and empty in all directions. Rather than cut the line and go adrift with the current, I figured the best place to be was at a known location. Quite a while passed before all the divers hanging on the anchor line completed their decompression, and the boat returned to the wreck site as the most likely place to begin searching for the only one missing. As Trzaska maneuvered the boat for the pickup, van Name leaned over the side and said with a grin, "It's a good thing for you we had to come back for the buoy."

Helm Stand Blooper

Not every mischance was resolved with such facility. A prime example was my introduction to the "Terror Wreck," an unidentified freighter off the border of Delaware and Maryland. The wreck was named by Bill Tattersall, captain of the *Selma* out of Indian River Inlet, Delaware, when Phil Lindale came up from the discovery dive with - as Tattersall described it - terror in his eyes. Lindale had never been down as deep as 170 feet.

I went to great lengths to dive on newly discovered wrecks, so a short time later I made the pilgrimage to Indian River. Tattersall hooked the wreck about midships. I wasn't long in the water when I found two portholes lying loose side by side. I sent them up on a liftbag that I tied to my decompression line, the other end of which I tied into the wreck after the liftbag reached the surface, so the portholes would not drift away with the current.

Almost immediately afterward I spotted a bronze helm stand. The wooden wheel had long since rotted away but the pedestal appeared undamaged. Hoping that the rudder indicator was imprinted with the ship's manufacturer's name, and would lead to the vessel's identity as

1970's: Decompression Comes of Age - 283

it had with the *Ioannis P. Goulandris*, I dragged the awkward assembly free of the wreckage - no easy task as I had to work ten feet of shafting out of overlapping beams - and eagerly attached my second liftbag to the base. When I thought about it, though, I calculated that the 200-pound liftbag lacked the capacity to float so heavy an item. It was also late in the dive. I was low on air and needed to save some for decompression.

Letting caution be my guide, I put only enough air in the liftbag to hold it aloft, then tied what remained of the decompression line to the pedestal and unreeled myself toward the surface. Since I always carried in excess of 350 feet of sisal, I planned to swim the reel to the boat after completing decompression, then use the line on the second dive to go directly to the pedestal. The plan would have worked had it not been for blind intervention.

I was decompressing comfortably at ten feet without a care in the world. My buoyancy was adjusted perfectly and the current was mild to middling. The water was clear as crystal. Jon Hulburt was having some difficulty. He was hanging nearby under a liftbag that was tied to the wreck. A steady stream of bubbles rose from a rent in the material. As the liftbag lost buoyancy, the weight of the porthole suspended beneath it pulled the liftbag under the surface. Whenever the liftbag sank within reach, he purged air from his regulator into the filler opening and sent the liftbag back up. Finally, in order to prevent the liftbag from taking a sudden nose dive for the bottom, he pulled some sisal off his decompression reel and proceeded to tie his sagging safety line to the one from which my portholes were suspended.

While I was watching his shenanigans I happened to notice my line going limp. I pulled up on it until I held a horizontal length between my hands. I glanced down and saw a rushing yellow warhead. What at first appeared to be a missile launched from a nuclear submarine was in actuality a pair of liftbags surrounded by a huge mass of bubbles formed by expanding air that was escaping voluminously from the filler openings. This self-propelled juggernaut was soaring skyward with only one obstacle in its path. Me. I kicked sideways frantically to avoid being struck as the liftbags and whatever they were hauling soared by.

The liftbags burst through the interface in a froth of foam. The surface was so flat and clear that I observed concentric shock waves spreading outward. As the agglomeration settled back down in the water and the bubbles slowly evaporated, the distorted image slowly resolved itself into recognizable context. The 500-pound liftbag overshadowed the 200-pounder bouncing off its side. From two lanyards dangled a monstrous bronze helm stand. The coincidence of a wreck having two nearly identical artifacts was just formulating in my mind when I noticed the ten-foot shaft protruding from the base of the stand. That was *really* odd.

A length of sisal extended down from the pedestal and disappeared into the gloom. At first I supposed this was a safety line that secured the pedestal to the wreck. But as the helm stand and I drifted away from Hulburt and my liftbag with the portholes, the light began to dawn on me: the decompression line that I held in my hands was no longer tied to the bottom but to the helm stand floating above me. I'd been cut adrift! I scrambled madly for the safety line that prevented my portholes from sailing into the sunset.

By this time Hulburt had completed securing a traverse line to his sinking liftbag, had severed his down line, and was pulling himself toward my new position of safety. I held the porthole line in one hand and my reel in the other. The helm was drifting away gently. I looped the reel several times around the vertical line, stuck the reel between my legs, cut what little line was left on the reel, and started tying knots.

Hulburt took in the situation at a glance. As an extra precaution against losing the helm stand, he worked his way back to his unstable liftbag and tied another line from that point to the line between me and the helm. What is short in the telling was long in the doing. By the time all these lines were laid and lashed, we had created an architectural nightmare that looked like a cat's cradle that failed, or a three-dimensional cobweb designed by a deranged spider. As the line to the helm stand stretched taut, the only vertical line still connected to the wreck began to take up the strain, and the liftbag supporting my two portholes was put to the test. The 100-pound liftbag was not only floating two portholes, but it was holding up against the horizontal stress applied by Hulburt's porthole and the helm stand.

Just about the time that everything was stabilized, decompression was over. I was out of air, out of line, out of liftbags, and out of temper. I was somewhat mollified when I reached the ladder and learned that I didn't have to chase after the helm stand. It was already secured to the boat with a stout length of rope. After doffing my tanks I swam upcurrent to my safety line, cut it, and drifted back to the boat. Plenty of help was on hand to retrieve the liftbags with the portholes, and to pull in the loose line leading to the helm stand. The pile of sisal looped on the deck was big enough to stuff a bushel basket.

Then I learned who the culprit was: Mike Boring admitted his guilt. He said he must have been narked out of his mind because he didn't notice the liftbag or my decompression line tied to the pedestal. He clipped on a 500-pound liftbag and sent the whole works to the surface. We all had a pretty good laugh about it.

For me, the day ended on a less than savory note. Calm seas yielded to a short chop that made suey in my stomach. My head was in a vice. But I didn't want to miss the second dive, on the *Nina*, a tugboat that foundered in 1910. On the bottom at 90 feet my condition grew worse. I recognized the symptoms: I was seasick. After twenty-five minutes I returned to the anchor line for ten minutes of decompression.

Trueman Seamans came up the line behind me. He saw that I was surrounded by a large cloud of silt, but couldn't see what my mesh bag contained to make such a mess. He cut through the cloud and peered through the mesh, found nothing inside the bag, made eye contact, then gagged and backpedaled in a frenzy when he saw me convulsing, and realized that the particulate matter in the water was my lunch.

The rudder indicator bore no inscription. The "Terror Wreck" remains unidentified.

The Deadeye Caper

Another indigestible dive occurred on the remains of a sailing ship called the "Benson." I knew of a deadeye that was secured to the wreck by metal rigging, so I went prepared with a hacksaw to cut the thing free. When I dropped down next to the grapnel to set it firmly in the wreckage, a sudden lurch caught me off guard. My hand got pinned under one of the tines. Luckily the grapnel slid off my palm and dug into my mitt. I wasn't hurt, but neither could I get my hand free. I thought I might have to pull my hand out of the mitt, and several anxious moments passed while I waited for the surge to release some slack, but I finally got free.

When I looked around to get my bearings, I discovered that my mesh bag was open and the hacksaw was gone. The whole purpose of the dive would be frustrated without the bladed tool, so I took a sighting on the anchor line and swam under it away from the wreck, figuring to come across the hacksaw which must have fallen straight down during my descent. It didn't work out that way.

I was out of sight of the wreck when I came across a different deadeye lying in the sand. I tugged, and it pulled loose. The wood was perfectly preserved, and the iron band that was bent around the perimeter ended in a jagged piece of chainplate that protruded half a foot from the lower edge. A strong current was running on the surface so I didn't want to send it up on a liftbag untethered. But it was so heavy that I couldn't swim with it. Consequently, I cradled it in my arms as if it were a baby and walked backward toward the wreck, like a hard-hat diver of yesteryear.

I must have veered at an angle and gotten off course. Suddenly my upper leg was struck as if by the jaws of a giant shark. It didn't bother me in the least because I knew it had to be Bart Malone trying to scare me. When I looked down, however, I couldn't see my leg. It had been swallowed from crotch to below the knee by the biggest goosefish

I've ever seen. A yardstick wouldn't have measured its width and it was nearly twice as long. The goosefish is an angler and one of the ugliest fish alive, with a round flat body covered with skin more warty than the wartiest toad that ever hopped. Its nom de plume is all-mouth, the reason for which was obvious. In restaurants and super-markets it's sold under the more catholic name of monkfish.

No matter what it's called, this particular one was clamped on my thigh with a grip like that of a vise. Furthermore, it was shaking me violently the way a terrier worries a rat. Without giving it a moment's thought, I raised the iron-banded deadeye high and swung it down on the all-mouth's head like a bludgeon. My piscine attacker instantly went bleary eyed, yawned its cavernous mouth, and slumped stupidly to the sand like a drunk after an all-night bender.

A jet of liquid spurted into my drysuit at the bend in the crotch. I prayed that it was a puncture in the neoprene rather than an unsa-vory lack of discipline. Now I looked around and saw that the bottom was blanketed with giant goosefish, as if I had stumbled onto a school outing. I navigated through the fish-filled obstacle course by looking over my shoulder and shuffling backward toward the wreck. I tripped over the turn of the bilge, humped the deadeye to the center of the sanded-in hull, then followed the keelson to the anchor line.

It was a simple matter to attach a liftbag and my decompression line to the deadeye, inflate the bag, and tie the line to a wooden beam near the grapnel. With my remaining bottom time I caught eight lob-sters. I decompressed on my safety line and hung under the deadeye, so I could cut the line when I was done and tow the liftbag to the boat. A shadow caused me to look up in time to see a diver kicking along the surface. I thought his purpose was to retrieve my liftbag. On the con-trary, he was exhausted and near drowning, and he climbed on top of the liftbag as if it were a life ring. I ended up towing him and the lift-bag to the boat. I never did find the hacksaw, but I didn't much care.

DUMB LUCK RECOVERY

Another dive that I botched initially but concluded with towering success - to the great annoyance of my friends - occurred when I dropped my camera in the water (shades of Mike de Camp) and went to look for it. Someone handed the rig to me as I was floating along-side the boat - my customary procedure. I thought I slipped the lan-yard over my wrist, but I must have missed, and I didn't notice right away because of the thick layer of neoprene. When I swam toward the anchor line I should have felt a tug on my arm, but I didn't. In a panic over the financial loss, I surmised what must have happened. I deflat-ed my drysuit and plummeted straight to the bottom in the hope that, without appreciable current, the camera would have preceded me on the same vertical course.

The camera was not visible within the thirty foot radius of ambient light. I pirouetted on my fin tips, and as I spun around looking desperately in the distance, I tripped over something between my feet. By incredible good fortune I was straddling the camera and strobe. This time I made sure that the lanyard was secured.

Rather than ascend and re-descend on the anchor line, I took a compass bearing and headed straight for the wreck. After swimming farther than what I judged to be the scope of the anchor line, I figured I must have miscalculated the distance. But when the wreck didn't appear after I doubled the reckoned distance, I couldn't overlook the fact that something was mysteriously wrong.

I rose to the surface to find myself a hundred yards behind the boat. I kicked with a will, but despite the mild current I was soon exhausted and out of breath. Joy Meredith let out more trail line. Joyously did I grasp the safety float. I kicked feebly, but I reached the boat more by dint of Meredith's arms than by my own legs. I stood on the bottom rung of the ladder for fully five minutes, gasping for air, before I felt recovered enough to continue the dive.

A few minutes later I was digging in the wreckage when I uncovered a brass chronometer in its original wooden box. Since I consumed most of my air during my wrong-way wanderings, I barely had time to compose a couple of photos and pull the chronometer and box from its bed of mud before it was time to go. The prize artifact met with jeers on the boat when my friends learned how I bungled the dive by following the wrong end of the compass needle. They felt - perhaps justifiably - that I didn't deserve a reward for such flagrant incompetence. They were more agitated two weeks later when I returned to the spot and found the wind-up key with the same serial number.

Mud Hole Misadventure

Not all my blunders eventuated fortuitously. In my opinion, the most hazardous wreck on the east coast is the *Choapa*, and this was truer in the 1970's before divers slashed away much of the monofilament that draped the wreck like a shroud. I've often said, "If you've dived the *Choapa* you can dive the *Doria*, but just because you've dived the *Doria* doesn't mean you're ready for the *Choapa*." I still maintain this axiom.

I was diving alone when I dropped over the back of the wheelhouse and became hopelessly entangled in fishing line of heavy-duty test. Thick strands of monofilament gripped me from behind, holding me as tight as a fly in a spider's web. No matter how hard I slashed over my shoulders with my knife, I couldn't cut myself loose. Either the blade was not reaching the strands or it simply pushed them aside instead of cutting through. I've always found it remarkable how short a time there is between recognition of danger and full-blown panic. I broke a record in reaching the latter state.

It was black as a tomb on the bottom, and the limit of visibility with a light was less than a body length (an appropriate analogy under the circumstances). I was about to slip out of my tank harness in a last-ditch effort to observe my bonds when my light happened to illuminate a ladder rung on the smokestack in front of me. At first I grabbed the rung only to steady myself, but when I saw another rung above it, I pulled myself up in order to make the lines taut for cutting. One of the older strands parted with a twang. I climbed to the next rung, and the next. Monofilament snapped off my tanks with each step up the ladder. Breaking strands sounded like the cracks of a bull whip. By the time I reached the top of the stack I was free of entanglement.

I was sorely shaken. I slid down the front of the stack to the bridge deck, oriented my way across the dark debris to the anchor line, and made my ascent. After decompressing, I climbed onto the ladder of the *Sea Lion* still spent from my ordeal. George Hoffman took one look at my face and knew that I had been through a terrible trial. He said that I was as white as a ghost, and that's exactly how I felt. My tanks and fin straps retained enough monofilament cuttings - streaming off behind me - to go several times around the boat.

DIVE BOAT ANTICS AND *SEA LION* SHENANIGANS

Trips with Hoffman were always a gamble. Invariably, if we wanted to go to a wreck that lay south of the inlet, he wanted to go north, and vice versa, despite calm seas and good weather. When I chartered the boat for The Dive Shop of New Jersey, I wouldn't stand for these idiosyncrasies that Roach was willing to accept out of peer pressure. Lichtman printed a schedule of dates and destinations, and people signed up for trips according to the wrecks they wanted to dive. All too often, Hoffman tried to intimidate me or Malone into going where *he* wanted to go. If we failed to remain firm in our resolve, we had to deal with unhappy customers who were, after all, paying for a service.

The *Sea Lion* was no longer the only boat in town, but there was no doubt in my mind that Hoffman was the best at finding wrecks. He could see a bump on a depth recorder that was invisible to me. He had lists of loran numbers from fishing boat captains, and often he would drop me off on a buoy line to check out a new site. Lobsters abounded on these low-relief snags because divers had never been there before. Hoffman kept these locations a closely guarded secret. He became livid with rage if he found another boat on a wreck that he considered to be "his." At these times he was so out of control that he spit and foamed at the mouth. He ate Maalox like candy. He seemed to think that the ocean and all its shipwrecks belonged to him.

Once on the *White Star* we dived a wreck that Captain Ray Ettel called the "Bonanza." It was well known to the boats that operated out of Barnegat Light. We were already anchored and had divers in the water when who should we see approaching from the north but the *Sea Lion*. When the boat got close, Hoffman gave the helm to one of the mates and ran up to the bow in a frenzy. At first I thought he was dancing a jig, but it soon became apparent that his wild gesticulations demonstrated anything but glee. He flung his arms in the air like a madman fighting off a swarm of hornets. He didn't need a radio. We could clearly hear every oath and epithet. He wanted to know how Ettel had found out about the wreck. Ettel shrugged, and told him that he fished and dived here all the time. The wreck was well known to the boats that ran out of Barnegat Inlet. It never occurred to Hoffman that a wreck that was a long way from his base of operations was close to other inlets.

Hoffman shouted across the water, "What do you call it?"

With dry aplomb and evident delight, Ettel cupped his hands to his mouth and called back, "George's Secret Wreck."

Hoffman had other quirks. I chartered the *Sea Lion*, either through the shop or on my own, to search for unknown wrecks. I got little or no cooperation from the diving community in this quest because people wanted a sure bet. If they put up their money for a day's diving, they wanted to make sure they got their two requisite dives. No one wanted to spend a day just looking. On the other hand, my philosophy was grounded on the assumption that the time spent searching for unknown wrecks paid off in handsome dividends. There was new territory to explore and loose artifacts to recover. One good dive could be far more rewarding than ten ordinary dives, and therefore was worth the time and money invested in the search. The most I could ever convince people to do was to spend half a day searching at the sacrifice of one dive. And even then they grumbled.

In my mind, the most efficient way to search for unknown wrecks was to utilize the surface interval between dives to check out coordinates in the area between one dive site and the next. That way, people got two dives for the day whether we found a new wreck or not. But Hoffman wouldn't go for it. He insisted that people relinquish a dive as a penalty for conducting a search.

I butted heads with Hoffman much harder on a different matter. On all but the deep dives, it was his policy to put a mate in the water to tie in the hook. Ostensibly this was to ensure that the boat did not break away while the customers were under water. In practice, however, he wouldn't let anyone else go down until the mate scoured the wreck for lobsters. His mates were such experts at grabbing bugs that they pretty much cleaned off the wreck before the customers had a chance. Off the Jersey coast, catching lobsters was the primary motivation for diving shipwrecks. Now the customers descended to the wreck to find nothing but shorts (bugs that were smaller than the legal size) and lobsters that huddled so far back in deep holes that they were impossible to reach.

Hoffman had arrangements with local restaurants to purchase the *Sea Lion's* catch. The money was then split among Hoffman and the mates for the day. Should any customers return empty-handed and brokenhearted, Hoffman, before sending his haul to market, offered to sell his lobsters to the very people he had bilked them out of. I stated in non-negotiable terms that on Dive Shop dives this unconscionable practice would cease.

Hoffman countered cleverly by invoking the catch-all clause that is used too often to disguise ulterior motives: safety. There was less chance of the boat breaking free if his mates tied in the grapnel. I thwarted this ruse by allowing that the mates could set the hook as long as they came right back up or did not catch any lobsters until after the customers had their turn. He objected, citing as his reason the extra effort required by the mate to ascend and redescend. I explained that the Dive Shop chartered the *Sea Lion* not for the convenience of the mates but to provide a service for the customers who were paying the bill. Ultimately, I overcame all his objections by proposing to set the hook myself or have Malone do it. We established that policy on all the boats, with one of us tying in and the other untying. Nor did we take lobsters until the customers had their shot.

When things didn't go the way Hoffman wanted them to he was wont to throw a tantrum. By tantrum I mean a display of temper that is stereotypically ascribed to a spoiled child whose unreasonable demands have been thwarted. When Hoffman had a fit of anger he screamed at people unlucky enough to be in close proximity, he flung his hands in the air like an off-balance tightrope walker, he kicked or tossed about the personal possessions of the customers (once he hurled my camera rig from one side of the boat to the other), and sometimes he spit so much that he appeared to be foaming at the mouth. This behavior contrasted sharply against acts of apparent benevolence, such as when he gave to a customer an artifact which had no value to him, or when he cheerfully offered to take a charter to a wreck that he wanted to dive. Dealing with Hoffman was difficult at best, and he and I often came to loggerheads, much like Captain Bligh and Fletcher Christian.

Perhaps Hoffman's most serious transgression was his refusal to cancel a charter despite gale warnings and Coast Guard small craft advisories. He wanted his money and he didn't care who suffered for it. Once I refused to load my gear on the boat because of the horrendous conditions at sea, preferring instead to pay for my spot and spend a comfortable day at home. Roach kowtowed to Hoffman's intimidation, but had to return my money when the *Sea Lion* was turned back by tremendous waves that nearly swamped the boat. Hoffman invoked the safety standard only when it was to his benefit to do so.

His other trick was to leave the marina under marginal conditions, then announce that the weather was too bad to leave the protection of shore, or predict that it might turn worse, in which case he dropped anchor on what he called the "Money Wreck." The "Money Wreck" was the *Delaware*, a small, wooden-hulled coastal passenger-freighter that burned to the waterline in 1898 some three miles from the inlet. Hoffman called the *Delaware* the "Money Wreck" because he made so much money on the site by taking people there when it was too dangerous to go any farther - and, in most cases, when it was dangerous even to be there. The *Delaware* was an interesting novice dive under suitable conditions, but when raging seas churned the shallow bottom into sandy particulate soup, visibility was generally nil, so that those hardy souls who plunged through the waves to reach the wreck found themselves groping blindly through suspended silt.

By taking the fifteen minute ride to the *Delaware*, Hoffman felt justified in collecting his full charter fee. By contrast, in the sport fishing charter business, when a captain can't take his people offshore to troll for marlin, he doesn't take them into the back bay to jig for perch. Malone and I, and The Dive Shop of New Jersey, felt an obligation to the customers, not only for their safety but for their recreational enjoyment. After all, wreck-diving was supposed to be fun, not a death defying feat.

Because of Hoffman's attitude in this regard, I made special arrangements with him to charter the *Sea Lion* out of season (early spring and late autumn) when the water was cold and when the boat sat idle at the dock, solely to dive deep in the Mud Hole. The coterie of deep wreck divers still numbered only in the handfuls - perhaps a score or more who were avid in the pursuit, and a few who dabbled. Ironically, this total was higher than the small fraction of EDA's membership who had dared extreme depths and who had enjoyed greater renown due to the chest-pounding stance of its president.

But as the 1970's drew to a close and a new decade dawned, deep wreck-diving teetered on the verge of a deeper plunge into more consistent waters.

Personal Notes

Meanwhile, I struggled to overcome personal bitterness in a society that did not live up to my expectations. I still sought the *Father Knows Best* fantasy land. When I returned home from war to so-called civilization, I believed in the sanctity of human endeavor. Yet whenever I took for granted that adults would behave with maturity, that the system of mores would triumph over transgression, that the law embodied justice, that reason would prevail, I was sorely disappointed. Rationality seemed to have no place in this land of unreason, where exploitation at the expense of others was the norm. My naiveté and my refusal to accept moral corruption in any form created depression and mental anguish.

The hard physical activity of construction work did much to strengthen and harden me physically, but it did not help overcome the cardiovascular disability of a damaged lung and compression muscles that were atrophied through nerve damage. Every inhalation felt like the third breath from a balloon. A fifty-foot dash left me gasping for air. Under water, I was constantly overstressed and always on the verge of overbreathing the regulator. On deep dives and during long decompressions I drained my tanks beyond the safety margin. So I took up jogging.

Not the popular activity then that it is today, I attracted queer looks from people I met or passed on the sidewalks. I also inhaled a lot of carbon monoxide, especially on hills. So I took to jogging in Pennypack Park, along the trails that I hiked and biked as a Boy Scout. At first I couldn't run more than a hundred yards before becoming completely exhausted by the shortness of breath - like an asthmatic in a hay loft. But I kept at it, once or twice a week and painful though it was.

At the end of a year I could jog half a mile at a steady pace. After two years I could go three miles by alternately running and walking. Then I was able to do the three-mile loop without stopping. Eventually I worked my way up to a loop that was six and a half miles long, which I continue to run today. My air consumption decreased by a third. I still consumed more air on a dive than anyone else in the world, but gauged against my previous consumption rate the improvement was pronounced. I felt nearly normal.

I took great pride in and received satisfaction from my work. Every day on the job was an accomplishment. I could look back over my daily labors and be rewarded by the result of productivity. Yet despite being a skilled and successful electrician, I was a failure as a union member. I was a constant thorn in the side of Local Union #98, International Brotherhood of Electrical Workers.

Anyone who has read *The Jungle* by Upton Sinclair knows what the world was like before the advent of unions. My paternal grandfather was laid off from his job with the railroad after nineteen and a half years so the company wouldn't have to pay retirement benefits. Unionism put an end to sweat shops, low wages, long hours, and shameful exploitation of workers by unscrupulous employers, in addi-

1970's: Decompression Comes of Age - 293

tion to setting standards of pay and fair treatment for non-union shops.

But unionism is a double-edged sword whose blade cuts both ways. Unions can be as corrupt as the companies whose policies they seek to temper. In the 1970's, the balance of power between contractors and unions swung like a pendulum out of control to the side of unionism, corrupting an inherently imperfect organization which operated beyond the pale of checks and balances that form the cornerstone of any democratic system.

A common union affectation in its opposition to the supervisory power wielded by contractors was the sanctimonious attitude of the membership with respect to perceived grievances. Unions stood up for members under the theory that they were innocent and could never be proven guilty. No matter what the circumstances, the contractor was always wrong. The epitome of this absurd attitude occurred when police investigators arrested a handful of union members who destroyed hundreds of thousands of dollars worth of property belonging to a non-union contractor. The local unions established a defense fund by increasing dues and by withholding money from our pay, with full knowledge that the accused members admitted to committing the crime. The rationale was that it was for the good of unionism.

I objected strongly to supporting the defense of acknowledged criminals. If they did something wrong, they should pay for it. The willful destruction of property stretched far beyond the presumption of upholding solidarity. My standpoint aroused disapproval from fellow workers who sided with the accused simply because they belonged to a union, as if membership bequeathed rights to which other citizens were not entitled.

I dealt with other union inequities on a more daily and more personal basis. The stereotypic blue collar worker is perceived as a person who guzzles coffee, gawks at girls, curses with regularity, and makes too much money. It's true that construction workers are well paid - and with good reason. Jobs are seasonal with unemployment common in the winter. The work can be harsh under outdoor conditions of extreme heat in summer and subfreezing temperatures in winter. And the work is dangerous.

This latter condition is often overlooked by white collar workers whose severest injury on the job might be a paper cut. Construction sites are full of hazards. Minor injuries abound, major accidents occur, death sometimes eventuates. I once had all the skin burned off my hands by hot tar that slopped through an air vent in the roof. During the construction of one skyscraper there were five fatalities in the eighteen months I spent on the job: a stone mason fell off the side of the building when a marble slab slipped from the clutches of a crane and fell through the roof of the cab, crushing the operator; one carpenter was catapulted out the side of the building (before the walls were erected) when the scaffold on which he was standing collapsed; another carpenter fell down an elevator shaft when he tried to walk off with a sheet of plywood that covered the opening; I don't remember the fifth.

An electrician was electrocuted when he grabbed a high-voltage bus bar that was mistakenly energized. His heart stopped and he was stuck to the copper bus by reflexive muscle action, but his partner kicked the ladder out from under him, administered CPR and mouth-to-mouth resuscitation, and brought him back to life. These accidents were deeply regretted, but no one thought the number was unduly high.

The truth of the matter is that most construction workers are conscientious, highly skilled, and honest - but they don't get their share of the publicity. Furthermore, the paranoia rampant in union ideology plays upon a person's innate weaknesses and subordinates his ability to overrule ingrained allegiance. Bad influence is difficult to overcome without uncommon strength of character.

I worked with some men who stumbled over an hour's work in a day only because it was unavoidable. Yet if a contractor laid off such a sluggard, the union took the side of the member despite the evidence against him. If unions had a mechanism for getting rid of idlers who give unionism a bad name, a greater purpose would be served. I knew for sure that our union was in trouble was when one of the worst loafers was elected business agent.

One foreman threatened to fire me because he said that my productivity made the other workers look bad by comparison. Yet, when I was foreman, I found that a hard-work policy can become infectious, resulting in greater satisfaction among the employees and more work accomplished for the contractor. The latter equates to larger profits, which does more to further the cause of unionism than complaining about picayune peeves.

Once when I was foreman we ran into a hang-up on the primary underground cable installation. While two men fed a bundle of high-voltage cables into a pipe from inside a manhole, and another soaped the feeder head to provide glide, three of us pulled on the rope at the other end a hundred feet away. The contractor showed up just as the cable jammed at the final elbow five feet from the junction box. No amount of tugging could pull it around the bend. "What are you going to do?" he asked with concern. I glanced at my watch, saw the time was noon, and said, "Break for lunch." He nearly threw a fit as I escorted the men to a nearby diner. Rested and refreshed, we returned half an hour later and completed the pull with little effort.

On opening day of a restaurant I spent the morning busily engaged in wiring the emergency lighting system while my co-workers completed other non-essential details "behind the scenes." Advanced advertising attracted people in droves. The owner was gratefully serving a full house. I made a final connection in the basement, screwed down the lid on the junction box - and pinched a wire between the sharp metal edges. The short circuit erupted in a ball of light the size of a grapefruit, nearly knocked me off the ladder, and plunged the room into blackness.

I felt my way up the stairs into the kitchen. A creepy feeling passed through me when I saw the cooks working the grills and ranges in the dim light of battery operated lamps. More than one circuit had been interrupted. I charged into the dining room where the secondary distribution panel was located, and was astonished to find the customers chewing and chatting in semidarkness. None of the breakers was tripped, not even the main. I removed the panel cover to put a voltage tester across the bus bars; no power was coming in. It was one of those rare instances in which the main feed breaker in the back room of the basement took the jolt before the secondary circuit breaker had time to trip. The whole floor was without light.

Just then the owner materialized behind me in a panic. The place was packed, business was brisk, and he was afraid it would all collapse around his ears. He squinted through disappearing dollar signs. "What's wrong?"

I glanced calmly across the dining room. Waitresses were serving, customers were eating, and no one seemed to be bothered by the shadowy veil that was broken only by the faint incandescents that had switched on automatically when the power went out. Indeed, I thought the ambiance had a somewhat romantic appeal. I looked him straight in the eyes. "I'm testing the emergency lights," I said. I turned on my heel and left him gaping.

Despite the camaraderie I shared with friends and co-workers, philosophical differences sometimes drove a wedge between relationships. What separated me the most from other people was a dissimilarity in world views and the comparison of human values. I opposed common unilateral thinking with my own brand of universal thinking. Whereas people generally live life in a conceptual maze, like lab animals whose vision is blocked by nearby partitions and whose outlook is limited by their frantic efforts to obtain self-gratification, my perspective rises above the maze, to a satellite position, from which I can look down and see my place with respect to my surroundings and judge how my actions might affect others.

This corny and self-defeating attitude was manifested dramatically when I refused to walk a picket line around a non-union job. According to union diatribe, non-union workers were "scabs" who took work away from the union. Try as I might, I was too blind to see any distinctions between union workers and non-union workers. All had families to support, mortgages to pay, children to love, and community obligations. My failure to distinguish one from the other led to arguments in which I defended the rights of non-union workers to earn a living and provide financial security for their families.

I reasoned that there was a discrepancy in the notion that non-union workers took away jobs. There were only a certain number of jobs to go around, and no matter who got those jobs, the same number of workers were employed. In the big picture there was no loss of work. In actuality, the distribution of work was uneven and discriminatory.

Instead of campaigning against non-union workers, I suggested, why not bring them all into the fold? Then "they" would be "we": one big happy family working together toward a more equitable lifestyle for everyone. This concept is no more socialistic than the basic doctrine of unionism. The majority of non-union workers would have jumped at the chance to join. The main reason they didn't already belong was because the unions wouldn't admit them.

I found no sympathetic ears among my fellow members. They all believed that the door to the local union hall should be shut behind them in order to protect the status quo. Thus unionism in practice was not the same as unionism in theory. In practice, unionism urged enforced inequality: disadvantaging certain individuals to sustain the advantages of the select.

This insufferable mindset reached idiotic proportions in an instance when fluorescent ceiling fixtures were delivered to a job from a factory that pre-wired the tube assemblies complete with a five-foot armored cable extension ready for splicing into a junction box. The union held that the cable extension took work away from electricians on the job. (In the past, the cable was cut and installed on site.) The union forced to issue by having a worker undo the internal splice, pull out the wires, hacksaw two inches off the steel cable to simulate cutting it off the coil, then refeeding the wires and making a new connection - on every one of thousands of fixtures. And the contractor had to pay wages for this absurdity!

I was working at the Philadelphia *Inquirer* when the newspaper's management decided to downsize the maintenance crew. My job was protected because the last one hired would be laid off first. No one could understand why I volunteered to be let go at a time when unemployment ran rampant throughout the trade, and when there was no hope of getting another job for months. I was only half sure myself that my contemplated career change was prudent.

The fact of the matter was that I wanted to be an author.

I had *always* wanted to be an author.

As a child I dabbled in science fiction. I read voluminously, daydreamed inveterately, and worked up stories in my head which I later committed to paper - solely for my own enjoyment.

When my tenth grade English teacher said we could earn extra credit by writing a term paper or short story, I accepted the assignment. My grades always needed an extra boost. I thought out one of my ideas, wrote a story in longhand, and asked my mother to type the completed manuscript. The result was forty-seven pages long, double-spaced. The teacher was so astounded he gave me an A for the course. Then I wrote a complete novel purely to give vent to my imaginative spirit. That cursive I never submitted.

After high school came college, war, marriage, and work - and no time to fulfill my literary ambitions. I needed money to support my family, and any spare time I managed to eke out between jobs and family obligations I spent diving and pursuing outdoor activities.

During my marriage, apprenticeship, divorce, and constant debt I got used to subsisting cheaply. Although my wife lived on a credit philosophy, I never spent money that I didn't have. On construction sites it was often my job to get coffee for the gang for the morning break. I took orders, collected money, and went to a local restaurant to buy perhaps a dozen coffees plus toast and doughnuts, then distributed the fare to wherever the men were working. I didn't have enough pocket change to buy coffee for myself, so I stoked up on individual creamers and downed the leftovers like shots of whiskey. On the rare occasions when a mechanic gave me extra money for a coffee for myself, or told me to keep the change, I made that worker first on my return rounds so he wouldn't know that instead of buying an extra coffee I had pocketed the coins. Perhaps there was some truth to my father's allegation that I kept my pretzel money in elementary school.

I never went out for lunch. I carried a lunch box with a couple of sandwiches and a book, and read by myself while my co-workers ate their noon repast at local bars and diners. I just didn't see any sense in spending money for the sake of deceptive appearances of affluence I found substance meaningful, and imagery as wasteful as vanity.

I still supported my son, but once free of a spendthrift wife I was able to save some money. My social activities were few to non-existent, I indulged in no extravagances, and although I lost my possessions and all the money I had ever earned and saved in my life, I gradually regained solvency. Eventually I had more money than I knew what to do with. So I read the *Wall Street Journal* and invested in the stock market. I didn't make a killing, but I made sound investments and overall I showed more profit than loss.

I bought an electric typewriter and a typing tutorial. By doing lessons nightly I taught myself to type. After gaining some proficiency on the keyboard, I kept my touch-typing in practice by typing notes from my shipwreck files and by writing research queries to libraries and museums. I also tried to do some creative writing but found that

I couldn't keep my mind on it: I was too tired from construction work to get my thoughts together for an evening session. I needed big blocks of time in which to concentrate.

This was my state of mind when the opportunity for a job dismissal presented itself. I was not only prepared to accept the offer, I was ambushing for it. With all that money in the bank and in the stock market, I could have bought a fancy new car, an ostentatious house, designer clothes, expensive vacations, and all the other affectations of conspicuous consumption. But I wanted none of that. I wanted to write, to create: a luxury that money couldn't buy.

Yet there was one important commodity that I found myself able to purchase, the one item that people always complained that they never ever had enough of.

Time.

I could use my savings to procure the time I needed to focus my attention on an occupation of my own choosing. Despite this rationalization and goal orientation, I felt strong trepidation over not knowing where my next dollar was coming from. I needn't have worried. I lived so conservatively that I was able to save money on the meager allowance provided by unemployment compensation.

Writing came hard to me. I admired authors with the facility to write long, fast, and well. For me it was a struggle. But despite the mental hardship of facing possible failure, what prompted me on my path was the fear of having regrets later in life over never having attempted to make my life what I wanted it to be. *That* would have been true failure.

Not having a regular job had other benefits as well. It enabled me to go on longer trips, either for diving or for outdoor activities, because I lost no wages from taking time away from work. I extended my underwater explorations to wrecks along the entire eastern seaboard, from Nova Scotia to Key West.

During this transitory phase I picked up odd jobs whenever work was available. New construction stood at an all-time low, with fully one-third of the union work force on the bench. For months at a time I was unemployed - that is, not gainfully employed: I labored all day at my typewriter. It was during this period that I moved to Boulder, Colorado, a quaint college town situated at the foot of the front range of the Rocky Mountains.

There were several reasons for the move at that time and to that location. I was deeply depressed after a failed love affair and was unable to keep my mind on my writing. The Denver-Boulder area was one of the few places in the country where there were more job openings than there were electricians to fill them. My local union was happy to give me a referral to the local union in Denver. Drew Maser, a fellow diver and shipwreck researcher, offered to rent my house for as long as I was on the road; he paid the utilities and the cost of the mortgage, thus relieving me of those financial responsibilities. And Paul Gafney was living in Boulder.

Gafney had forsaken wreck diving for a new passion: rock climbing. He phased out diving, climbed cliffs along the east coast, and finally decided that he loved the sport so much that he had to relocate to where greater challenges were presented. The Rockies seemed like an optimum place to climb. After he moved, we maintained our friendship through correspondence. When I decided that I needed a change in atmosphere, and found that there was work in the area, I called him and explained my situation. He invited me to stay with him until I found a place to rent.

Since I did a fair amount of rock and mountain climbing, and had made several trips west, getting my hands back in the electrical business as well as onto the rock ledges seemed like good medicine for a broken heart. I went so far as to contemplate a permanent change in goals and geography: quitting diving for good and staying in Colorado to embrace the outdoor life. I even sold some of my dive gear.

Gafney and I did some great climbs together, beginning with the Third Flatiron that towers above Boulder. I also made quite a few solo climbs, backpacking into the high country where I felt the freedom of solitude and enjoyed the denunciation of the daily grind. He and I had different degrees of skill and security. Whereas Gafney was a staunch believer in roping up for a climb, I disdained the use of ropes. In many situations in which he relied on a rope for safety, I found it an unnecessary encumbrance. I embraced the unfettered exhilaration of climbing unaided and unprotected. Because of this, there were some mountains that Gafney wouldn't climb with me. Those I climbed alone.

My most satisfying solo climb was the Crestone Needle, in the Sangre de Cristo range south of Denver. I backpacked to a secluded pond at an altitude of 12,000 feet. To the east I had an unobstructed view of the vast plains a mile below. The other three cardinal points were marked by fourteeners (mountains whose peaks rose above 14,000 feet). I didn't have time to climb Kit Carson, the peak to the west. One day I walked up the gradual slope to the northern peak, Mount Humboldt. The next I decided to climb the vertical face of the Crestone Needle, the more difficult of the pair of peaks that stood side by side, the adjacent one being Crestone Peak.

I didn't sleep in a tent but in the open air. I hung around my sleeping bag till lunch, reading, so I wouldn't have to carry a pack with food and water. I planned to make a fast assault without any equipment. The day was warm and sunny. I wore only jeans, tee shirt, socks, and jogging shoes. I hiked to the base of the mountain where the 2,000-foot ascent began up an ever-steepening rock wall.

The mountain was rugged and challenging and, despite many

ledges only a finger thickness in width, within free-style capability. Two hours of continuous climbing found me 1,500 feet above my starting point, with an awesome view of the col where my bright yellow sleeping bag cover contrasted sharply against the bright green grass and deep blue water of the pond. Then I heard voices on the pitch above. Two college students were lead-climbing the same route that I was free-climbing.

Lead climbing is a leap-frogging technique in which one climber scales a pitch "protected" by being roped to another climber who is secured to the mountain by means of a safety rope anchored by pitons, chocks, loops, or expansion devices. If the lead climber falls, he drops twice the distance to his last piece of protection, or, if he isn't installing protection, twice the distance to his safety climber. These guys were fully rigged, with hanging carabiners clattering like Chinese wind chimes, and dressed for the part, including specially made rock climbing boots.

I chatted with the safety climber for a couple of minutes until the top climber called down, "Off belay." Then I casually scrambled up the pitch alongside the rope, to the amazement of the top climber who was securing himself to an outcrop in order to provide protection for his companion. "Where's the rest of your party?" he asked.

"I'm alone."

He was shocked. Since I was new to the area I didn't know much of the local lore, but he assured me that the Crestone Needle was the most respected climb in Colorado. He offered to let me make the summit bid with them, but I declined because, in the course of conversation, I compared our rates of progress. They began the climb that morning at 8 o'clock. By their slow, plodding, and methodical procedure, they managed to climb in six hours what I accomplished in two. At the rate they were going they wouldn't reach the summit by nightfall. I had different designs.

I thanked him for the offer, wished him luck, then took off on my own. An hour later, when I climbed over the sheer face onto the pointed peak, two hikers were as shocked at my appearance as if I had climbed over the top-floor railing of the Empire State Building - which is only half as high as the Crestone Needle. After a friendly chat, I sped down the trail they had ascended on the back of the mountain. Long after dark, after a hot freeze-dried dinner and a couple cups of coffee, I was nestled cozily in my sleeping bag, reading by flashlight, when I noticed lights bobbing along the approach trail and a party of climbers descending to the col and on to the lower saddle. For them the Crestone Needle had been a thirteen hour ordeal.

The vaulted perceptions I beheld in the mountains made it possible for me to re-evaluate my life and what I wanted to accomplish with it. I loved Colorado and the splendor of the great outdoors, but I felt the lack of the many bonds of friendship that I had acquired in such depth, and I missed the excitement of underwater exploration. I have returned to the wilderness many times, but outdoor adventure was not a pastime that I wanted as a way of life.

I came home.

Maser and I shared accommodations until he got married. I got a job in downtown Philadelphia, doing renovations to large office buildings. But I knew for sure now that electrical work was not what I wanted to spend my life doing. Nor was I dazzled by the consumerist formula that defined success as the accumulation of material wealth. The American way was not my way. I didn't believe that a person's worth was based upon the value of his estate at the time of death. Power and possessions are delusions that die with the body. Only fools chased money as an end to all means. The more money, the bigger the fool.

I had a different calling. My calling.

I was more interested in what I could create than in what I could possess or whom I could control.

After a goal-setting session with my broker, I traded my portfolio from long-term growth stocks to high-yield utility stocks. Monthly dividends plus my disability pension made it possible for me to survive in a no-frills lifestyle, at a level of bare subsistence that most people would find constraining. But since I wasn't obsessed by the desire for creature comforts, and didn't need to be surrounded by sumptuous belongings, I was content with a simplicity of habitation that bordered on poverty.

But poverty is a point of view. Time, I hold, is a valuable asset.

I decided to go for broke. And so, at the age of thirty-two, I retired.

To Be Continued In. . .
The Lusitania Controversies
Dangerous Descents into Shipwrecks and Law.

The sequel volume continues the exciting story of wreck-diving, with deep penetrations into the *Andrea Doria*, mixed-gas explorations of the German battleship *Ostfriesland*, at 380 feet, a detailed account of the technical diving expedition to the *Lusitania* in 1994, and the court manipulations that followed. You won't want to miss it.

Index

Names in italics refer to ships, boats, books, songs, poems, newspapers, magazines, films, or television shows. Page numbers in italics refer to photos or captions.

abandoned houses: 118
Acton, Murial: 68
Adkins, Bruce and Rodney: 81-82
Admiral Farragut (school): 241
Admiralty: see "British Admiralty"
Alamo: 37
Alaska: 200
Albatross III: 91
Allies: 18, 21, 38, 43, 90
Already Salvage and Towing Syndicate: 51, 53
Amazon River (South America): 36
Amberjack II: 263
Ambrose Channel (New Jersey/New York): 12
American Littoral Society: 263
American Revolution: 128
America syndrome: 206
Anchorage (Alaska): 200
Andrea Doria: 64, 104-109, *174*, 256-263, 287, 301
Andrews Air Force Base: 200
Andros Island (Bahamas): 215
anti-German sentiment: 43, 45, 46
Apocalypse Now: 126
Aquadro, Surgeon-Commander John: 66-68
Aquamaster regulator: 106
Aquarama Dive Club: 218, 241
Aquarians (dive club): 270
Aquatic Recreational Enterprises: 271
Aquitania: 154
Arabic: 26, 38, 43
Archambault, Bob: *169*, *174*, 217, 237-238, 240-241, 250-253, 256
Arctic: 43
Ardolino, Fred: 263
Army Corps of Engineers: 12
Army of the Republic of Vietnam: 185
Arpione: 59
artifact:
　defined: 101-102
　displays: 269-271, 280-281
Artiglio: 59
Arundo: 102-103, 104, 106, 229-230
Aspinall, Butler: 33
Asqui, John: *174*, 256
Athenia: 36, 43
Atlantic ferry: 13
Atlantic Twin: 174, 256, 258
Atlantis Rangers: 104, 281
Atlas, Charles: 74
Ayuruoca: 99, 104, 236, 248
Axis powers: 69
Bacardi (wreck): see *Durley Chine*
Bahamas: 70, 215, 258, 273
Bailey, Ray: *174*, 256
Balaena: 270
Baptist: 79-80
Barnegat (New Jersey): 91
Barnegat Light (New Jersey): 263, 289
Bartram, Evelyn: 100, 109, 110
basic training: 129-131
Bass: 107-108, *169*, 230, 234-236, 272
Batman and Robin: 100
Battle Hymn of the Republic: 37
battle syndrome: 205-206
Beasley, Risdon: 62-64
Beaujean: 95
Belgian Red Cross: 51
Belgium: 39
bends: 51, 57, 67-68, 93, 223-226, 243-248
Benson (wreck): 285-286
Berkeley (California): 70
Berlin (Germany): 48

Index - 303

Bermuda: 62, 271
Bianculli, Joe: 242
Bible, Holy: 80
Bidevind: *174*
Bigham, Sir John: see "Lord Mersey"
Big Jim: 91, 110
Big John: see John Dudas
Black Point: 270
Black Sea (Europe): 40
Blakely: 51-54
Blavat, Jerry: 203
Bligh, Captain: 290
Block Island (Rhode Island): 106
Blue Hole (cave): *166*, 213, 214, 218
Blue Ribband: 12-13
body count: 144
Bogart, Humphrey: 65
Bonanza (wreck): 289
Boone, Daniel: 102
Boring, Mike: 284
Boston whaler: 97, 209
Bottom Time: 100
Boulder (Colorado): 298-299
Bowman's Tower (Pennsylvania): 119-125
Boyle, Mike: 103-104
Boy Scouts of America: 83, 130, 133, 292
brass fever: 101
Brazil: 99
Bremmer, Elizabeth Jane: 49
Bressette, Danny: *172, 174*, 217, 236, 245, 250, 256-258, 281-282
Brewer, Jack: 109
Bridges, Lloyd: 88
Brielle (New Jersey): 89, 97, 240, 256, 270
Britain: see "Great Britain"
Britannic: 26, 43
British Admiralty: 9-10, 16, 18, 22-23, 25, 31-32, 35, 40, 51-53, 58, 63, 65
  instructions: 33-35
British Board of Trade: 29-30, 35, 51, 58
British Isles: see "Great Britain"
Brow Head (Ireland): 24, 34
Brown, Jack: 92, 106, 108, 109
Brown, John (and Company): 9

Brunel, Isambard Kingdom: 9
Bryan, William Jennings: 37, 65
Bullock, Dave: 277-279
Burdewick, Ron: *174*, 256-259
California: 70
Californian: 30
Camburn, Richie: 204
Candidate: 23
Cape Clear (Ireland): 35
Cape May (New Jersey): 91
Captain Chum: 91, 110
Capt. Cramer: 277-278
Caribbean: 70, 129, 215-216
Carl D. Bradley: 43
Carr, Greg: 112
Carr, Mrs.: 112
Cassidy, Edward: 54-55
Catholic: 79
cave exploration: 210-214, 253
Cayo Romano: 22
Cedarville: 43
Central Powers: 36
Centurion: 23
Charles H. Morse: 95
Chee, Winston: 106
Cherokee: 97, 101, 241-243, 270
Choapa: 236, 287
Christian, Fletcher: 290
Churchill, Winston: 16, 31, 35-36, 40
Ciampi, Elgin: 70
City of Perth: 95
Clydebank (Great Britain): 9
Coast Guard: see "United States Coast Guard"
Cobh (Ireland): see Queenstown
codeine: 195
Coimbra: 97
Coleman, Carol: 264-265
Collector of the Port of New York: 65
Colorado: *168*, 298-300
Coningbeg lightship (Great Britain): 23-24, 35
Connecticut: 58
Constitution: 184
Continental soldiers: 128
cootie: 72
Coupe de Ville: 129
court-martial: 182, 184

Craig, John D.: 59-61
Craig-Nohl dress: 61
Cramer, Captain Don: 241, 278
Crestone Needle (Colorado): *168*, 299-300
Crestone Peak (Colorado): 299
Crilley, Frank: 57-58
Cub Scouts: 272
Cunard Steamship Company: 9-10, 16, 21, 38, 50, 65, 95
CY Divers: 218
Cylinder, Robert: 76
Daisy B.: 92
Danger Is My Business: 59-61
Daniel J. Morrell: 43
Daniels, Josephus: 37
Darmstadt (Germany): 47
Davis, G.H.: *156*
Davis Observation Chamber: 54, 59, 60
Davis, Robert: 54, 59
de Camp, Michael A.: 70-71, 89, 91-97, 100, 102-110, *165*, *171*, *174*, 228, 229, 276-277, 286
Decatur Road (Philadelphia): 114-118, 141
Decca: 91
Decker, Tad: 74
decompression: 61, 245, 271
    remote: 227-228
    teaching: 272-273
decompression meter: 223-224, 262
decompression reel
    invention: 106
    evolution: 226-228
Deep Diving and Submarine Operations: 54
Delaware: 291
Delaware River: 119, 121
Delaware Underwater Swim Club: 281
DelCorio, Lynn: 276-277
DellaBadia, Gene: 224-225, 253
DeLucchi, Chris: 257-259
Demerol: 191, 194, 196
Demetriades, H.J.: 59
Denmark: 236
Denver (Colorado): 298-299
de Page, Antoine: 51
dive master duties: 275-276

Dive Shop of New Jersey, The: 271-279, 281-282, 288-291
Doeblin, Alfred: 48
Dover (Great Britain): 54
Dow, Captain David: 16, 18-19, 31
draft (military conscription): 129
drag racing: 114-118
Droste, C.L.: 46
Dubeck, Al: *166*
Du Chef, Riviere (Quebec): *167*
Dudas, John: 100-102, 104, 106-110, *170*, 249-250, 258
Dulinski, Captain Jim: 91, 110
Dunn, Charles: 95
Dupont Employees Skin and Scuba Diving Club: 281
Durley Chine: 97
Dutch: 95
Dwyer, Donn: *174*, 256, 258
Earl of Latham: 22
Eastern Divers Association: 110, *174*, 208-209, 216-218, 223, 230, 231-234, 248-249, 262, 271, 274
    collapse: 267-269, 279-281
Eastland: 43
East River (New York): 43
echo location: 90
Edmund Fitzgerald: 43, 47
Eggesford: 238
Egypt: 59
Ellis, Captain Frederick D.: 45, 46
Empire State Building (New York): 300
Empress of Ireland: 25-26, 29-30, 43, 45
Engineering and Scientific Association of Ireland: 52
England: see "Great Britain"
English Channel: 19, 40
Enright, Jay: 112, 119-125, 200
Enright, Kenny: 112-113, 119-125, 200
Ensor, H: 52
Entler, Joel: *174*, 256, 270
escrow fund: 232-233
Ethiopia: 61
Ettel, Ray: *174*, 263, 289
Fastnet (Great Britain): 35
Fate of the Andrea Doria: 109

Index - 305

Father Knows Best: 210, 219, 291
ferry boat artifact exhibit: 240
Fire Island (New York): 263
Fischette, Captain Charlie: 95
Fisher, Admiral (First Sea Lord): 20
flag, neutral: 18
Ford and the Ferrari: *174*, 257
Ford, Dave: 270
Ford Galaxie: 113
Ford, Gary: 241-243, 250, 253, 270
Foreign Office: 58
Forsburg, Captain Paul: 91, 106, 108
Fort George G. Meade (Georgia): 200
Fort Hamilton (New York): 13
Fort Polk (Louisiana): 131-135
Fort Wadsworth (New York): 13
Four Tops: 203
Fox, Joseph: 105
France: 37, 59
frat: 112
Freeport (New York): 91
Frost, Wesley: 46-47
Fuhrer: 226
Gafney, Paul: 268-269, 280, 299
Galaxie, Ford: 113
Gallipoli Campaign: 41
Galluccio, Captain Joe: 89-90, 92, 104
Geelong: 53
General Electric Company: 60
General Fleischer: 99
General Slocum: 43
Gentile, Domenic and Meta: 71, 200
Gentile, Michael: 200, 219-220
George's Secret Wreck: 289
Georgia: 130
German codes: 31
German Submarine Warfare: 46-47
Germany: 16, 18-21, 24, 29-31, 35-38, 48, 49, 53, 55, 60, 65, 104
Gertie, the Trained Dinosaur: 47
Gimbel, Peter: 105
Glas, Brad: 234
Glasgow (Scotland): 59
G. Magnus, Law of: 103, 233
Gmitter, Tom: 85-87, *166*, 210
Godzilla: 195
Goetz, Karl: 38-39

gold shipments: 13
goosefish: 285-286
gorilla diver (defined): 208-209
Grand Dame of the Sea: 263
Grand Fleet: 16, 40
Great Britain: 12, 18-20, 29-31, 35-38, 43, 47, 55-56, 60, 61, 63, 236, 238
Great Eastern: 9-10
Great Northeast: 72-79
Great War: see "World War One"
Great White Fleet: 37
Greece: 59
Hackett, Eddie: 112-114
Hague Convention: 17-19, 38, 53
Hansel and Gretel: 264
Havana Harbor (Cuba): 26
Helen II: 234
helium: 61
Hepler, Paul: 270
Hesperus, The Wreck of the: 47
Hilsinger, Dick: 100, 102, 106, 109
hitchhiking: 204
Hitler, Adolf: 61
Hoehling, A.A. and Mary: 64
Hoffman, George: 97, 100, 102-104, 106-110, *170, 172-174*, 209, 217-218, 224, 244-245, 248-249, 270, 277, 288-291
Holland: 39, 57
Holman, Joe
Hoodiman, Bill: 107, *169, 172-174*, 217, 237-238
Hopkins, May Davies: 49
Horrors and Atrocities of the Great War: 44-45
Hudson Canyon: 98
Hudson River (New Jersey/New York): 22, 98
Hughes, Tommy: 14
Hulburt, Jon: 4, *169*, 228, 283-284
Humboldt, Mount (Colorado): 299
Hvoslef: 97, 101
hyphenated Americans: 45
Idicula, Dr.: 225-226
Imperial High Seas Fleet: 16, 40
Indian River Inlet (Delaware): 282
Inquirer, Philadelphia: 296
International Brotherhood of Electrical Workers: 292

306 – The *Lusitania* Controversies

International Salvage Court: 52
Interstate 95: 113-114
Ioannis P. Goulandris: 97, *169*, 236-238, 240-241, 244, 250, 270, 283
Ireland: 22-24, 46, 50, 64
Irish Channel: 35
Irish Free State: 50
Irish Sea: 18, 22-23, 31, 34
Israel: 95
Italia: 108
Jacob Jones: 101, 280
Jarratt, Jim: 60, 66
Jess-Lu: 91
jogging: 292
Jungle, The: 292
jungle warfare training: 131-135
Junior (wreck): 99, 236-238
Juno: 32
J-valve: 93
Kaiser: 21, 30, 36, 38-39, 63
Kaiser Wilhelm II: 12
Kamahoali Dive Club: 215, 217, 218, 223
Karloff, Boris: 193
Kaye, Danny: *164-165*, 197-198
Key West (Florida): 298
Kidde fire extinguisher: 106
Kinsale (Ireland): 68
Kishine Barracks (Japan): *164-165*
Kit Carson (mountain in Colorado): 299
Kiwi: 282
Klein, Howard: 67
Krasberg, Alan: 256
Krotee, Walter and Richard: 97
Krumbeck, Walt: 104, 107, 109, *174*
Lake-Railey Lusitania Expedition: 56-58
Lake, Simon: 56-58
Landi, Count: 54
land ranges: 89-90
Lansing, Robert: 37
last rights: 190
Last Voyage of the Lusitania, The: 64
Laurel Creek Cave: 211-213
Laurentic: 56
Lauriat, Charles: 45-46
Law of G. Magnus: see "G. Magnus, Law of"
Learning to Dive: 214
Leavitt, Benjamin: 51-54
Leavitt-Lusitania Salvage Company: 50-54
Leith, Robert: 28
Leopoldville: 43
Lichtman, Norman: *175*, 271-279, 281, 288
lifeboats: 23
Lightfoot, Gordon: 43, 47
Light, John: 65-69, *164-165*
Lillian: 97
Lindale, Phil: 282
Link, Helen: 214-215, 217
Little John: see John Pletnik
Little Nemo: 47
Liverpool: 13-14, 16, 18, 22, 34, *146*
Liverpool and London War Risks Insurance Association: 58-59
Lloyd Brasileiro: 99
Lloyd's of London: 57
Lloyd's Register of Shipping: 53-54
Local Union #98: 292
London (Great Britain): 29
Longfellow, Henry Wadsworth: 47
Long Island (New York): 91, 94-95, 99
Long's Peak (Colorado): *168*
loran: 90-91
Lord, Captain Stanley: 30
Lord, Walter: 42, 64
Lost World, The: 132
Louisiana: 131-135
Lusitania:
  books: 44-48
  career: 13-16
  conspiracy theory: 31-32, 35-36, 38-39
  construction: 9-10
  film: 47
  insurance: 21
  interior decor: 10-11
  launching: 11
  maiden voyage: 11-12
  mythology: 40-44
  photographs: *145-163, 176*
  propellers: 10, 13, 14, 15
  reparations: 48-49
  sinking: 26-28

Index - 307

stage: 47-48
storms: 14-15
torpedoed: 24-26
turbines: 10, 15
Lusitania Case, The: 46
Lusitania Salvage Company: 50-54
Lusitania's Last Voyage, The: 45-46
Lutine: 57-58
lymphatic edema: 225-226
Maalox: 289
MacLeish, Kenneth: 66-68
Magnus, Law of G.: see "G. Magnus, Law of"
Maharajah: 53
Maine: 26, 37
Main Line Divers: 218, 234, 268
Maleschewski, Ed: 100
Malone, Bart: *175*, 273-279, 285, 288, 290-291
Manasquan Inlet (New Jersey): 209
Mansfield, Mt. (New Hampshire): *168*
Marconi: 11
Marne, First Battle of the: 41
Marshall, Logan: 45, 46
Maser, Drew: 4, 298-299, 301
Mauretania: 9, 12-13, 16
Maurice Tracy: 217-218
Maxter Metals: 263
Mayflower: 16
McAllister, Joe: 78, 80
McAllister, May: 82
McAllister, Michael: 80-83
McCay, Windor: 47
McIlwee, Tom: 110, 227-228, 241-243
McNeil, Chief Officer: 14
Mediterranean Sea: 40, 53, 100
Meredith, Joy: 287
Mersey, Lord: 29-30, 32-35
Messina, Frank: 235, 242-243, 245
Michigan (Lake): 61
Mimosa: 18
Miss Shelterhaven: 241
Mistassini River (Quebec): *167*
Minnehaha: 59
Mixed Claims Commission: 48-49
Modernists: 48
Mohawk Canoe Club: 219
Mohawk (revenue cutter): 270

Money Wreck: 291
Montauk (New York): 91, 106, 109
Mont Blanc: 43
Moonstone: 97, 101, 110, *171*
morphine: 189-190, 196
Morristown (New Jersey): 70
Morse code: 11
Morse, Dennis: 66
Mount Fuji (Japan): 195
Mount Sinai Hospital (New York): 247
Mud Hole: 98-99, 208-209, 216, 236-238, 282, 287-288, 291
Mummy, The: 193
Murtha, Jim: *168*
Mussolini, Benito: 61
My Lai (Vietnam): 184
1984: 130
Nagle, Bill: *169*, 234-236
Nagrowski, Jan: *174*, 229-230, 256-258
Nantucket (Massachusetts): 105
Nash, Tom: 112
Nassau (Bahamas): 70
National Archives (U.S.): *147, 154, 155, 158-159*
National Broadcasting Company: 66
Nationalists: 48
Nazi: 39, 61
near-death experience: 188, 192, 254-255
neo-Nazi: 226
nerd: 76, 111
Neuman, Fred: 271
neutrality: 17-18
Newcastle-upon-Tyne (Great Britain): 12
New Jersey: 70-71, 89, 91-92, 94, 96-101
New Jersey Turnpike Commission: 282
Newport News Shipbuilding and Drydock Company: 49
New York Aquarium: 70
New York City: 14, 15, 21, 29, 43, 53, 95, 105, *155*, 247-248, 282
New York harbor: 13, 94, 98
New York Times: 54, 55-56, 99
Night to Remember, A: 42, 64

Nina: 284-285
nitrogen narcosis (defined): 216
Nitsch, Don: 107, *174*, 234-236, 246, 250, 253, 267
Nohl, Gene: 61
Norddeutscher Lloyd: 12
Northeast Philadelphia Airport: 114
Northern Pacific: 97, 101, 278
North Sea: 19-20, 38
Norway: 20, 95-96, 99
Nova Scotia: 298
106th General Hospital (Japan): *164-165*
Oceanographic Historical Research Society: 94-95
Ocean Venture: 103-104
O'Connor, Jerome: 50
Oil Wreck: see *Ayuruoca*
Old Head of Kinsale: 22, 24, 28, 33-34, 59, *148, 163, 176*
Oregon: 95
Orkney Islands (Scotland): 20
Orphir: 59-61
Orwell, George: 130
Osborne, Chuck: 66
Ostfriesland: 301
O'Sullivan, Paddy: *165*
out-of-body experience: 188, 192
oxygen poisoning: 61
Palmer, Robin: 106
Parcheesi: 36
Patane, Angelo: 275
Pennypack Creek: 76
Pennypack Park: 183, 292
Peress armored diving dress: 59-61
Peress, Joseph: 59-61
Petersfield: 104
Peterson, Gene: 4, *175*, 274, 276-277
Peterson, Joanie: 4
Philadelphia (Pennsylvania): 53, 71-79, 112-118
Philadelphia Civic Center: 240, 280
Philadelphia Maritime Museum: 240-241
Philippines: 194
Pinta: 95, 100, 111, 238-240, 248-254
Pletnik, John: 100, 106-107, *170*, 217, 229, 244-245, 248-255

Point Pleasant (New Jersey): 100
Poland: 61
pony bottle (invention): 106
Poponi, Dave: 276
Porter, Captain Jay: 91
post-traumatic stress syndrome: 205-206
Prater, Cal: 109
progressive penetration (defined): 264
Purple Heart: 193, 194
Purtle, Bill: 112-113, 119-125
Q-ship: 19
Quebec: *167*
Queenstown (Ireland): 11, 22, 28, 29, 32, 46, 50, *158-162*
Quonset hut: 193, 194
Railey, Captain H.H.: 56-58
rapture of the deep: 216
Ray, Carleton: 70
Receiver of Wreck: 58
recreational diver (defined): 208
redcoats: 128
Red Cross, American: 182, 185, 194, 197, 200, 201
Red Cross, Belgian: 51
Reese, Bill: 112, 119-125, 190, 201
Rhine (river in Europe): 61
Richards Army and Navy Store: 70
Rieseberg, Harry: 97
River of Doubt (South America): 36
Roach, Tom: *170-171, 174*, 209, 214, 216-218, 229-230-238, 243-250, 253, 256-258, 267-269, 277, 279-281, 288, 291
Robbins, Captain Chum: 91, 110
Robert Roger & Co.: 238
Roberts' Rules of Order: 29
Roberts, Smokey: 109
Rocky Mountains (Colorado): *168*, 298-299
Rocky Mountain spotted fever: 267
Rodocker, Don: 257-259
Roosevelt, President Teddy: 36-37, 45
Royal Navy: 62-64, 66
R.P. Resor: 97
Ruark, Walter and Pauline: 78-80, 182-183
Rush, Ed: 100, 102, 106, 109

Index - 309

Russian roulette: 94
S-5: 280
Sager, Harley: 276
Saliger Ship Salvage Corporation: 263
Salisbury (Maryland): 71, 78-83, 182
San Diego: 263-267
Sandy Hook (New Jersey): 22
Sangre de Cristo Mountains (Colorado): *168*, 299-300
Santa Claus: 79-80
Sauder Collection, Eric: 4, *145, 146, 148-150, 152-160*
Sawyer, Tom: 30
Scally, Frank: 106
Schieber, Jack: *168*, 219
Schieber, Rose Marie: 219
Schleinitz, Vizeadmiral Freiherr von: 21
Schwieger, Kapitanleutnant Walther: 22-26, 30, 37
Scientific American: *151, 153, 176*
Sea Hunt: 88
Sea Hunter: 256
Seal: 100
Sea Lion: *170, 172-173*, 209, 216-217, 230, 245, 248-249, 280, 281, 288-291
Seamans, Trueman: 285
Sea Ranger: 89-90, 104
search and destroy missions: 138-139
Second World War: see "World War Two"
Selma: 282
Semper Paratus: 54
sewers: 83-87
Shalom: 95
Shark River Inlet (New Jersey): 282
Sheehan, W.J.: 53
Sheepshead Bay (New York): 263
Shipwrecks off the New Jersey Coast: 97
Shukus, Captain John: 91
Siebe, Gorman & Company: 54
Sierra Madre: 101
Sinclair, Upton: 292
Sinking of the Lusitania, The: 47

Skagerrak (North Sea): 19
Smith, Captain E.J.: 30
Smith (Drew, Gwen, and Todd): *168*
Snediker, Graham: 95
Snivley's Cave (Maryland): *166*
Snyder, Jim: *174*, 256
Somme, Battles of the: 41
Sommerstad: 99-100
sonar: 90
Sorima Company: 59-61
SOS decompression meter: 223-224, 262
South America: 36
South China Sea: 126
spelunking: see "cave exploration"
Sphere, The: *156-157*
Stahl, Gustav: 65
Starace, John: 107, *172-174*, 256-262
Steuben: 43
St. Christopher's medal: 190, 194
St. George's Channel (Great Britain): 22-23, 34-35
St. Lawrence Seaway (Canada): 30, 43
Stockholm: 105
Stockton-on-Tees (Great Britain): 238
Stoker, Kenneth: 75
Stolt Dagali: 95-98, 100, 111, 232, 272
Stone Harbor (New Jersey): 91
Storey, Third Officer: 14
Storstad: 30
Stratton, Captain Charlie: 100
strict accountability: 36-37
Strykowski, Joe: 214
Subervi, Elliot: 110, 208-209
Swan Hunter and Wigham Richardson: 12
Sweden: 105
Tattersall, Bill: 282
Temple University: 129
Temptations: 203
Terror Wreck: 282-285
Testa, Frankie: 76-77
Texas Tower: 97, 230
Third Flatiron: 299
Three Daughters: 95
Thumper: 91

Tiger Land: 131-135
Timex watch: 93
Titanic (1953 film): 42, 64
Titanic (wreck): 15, 26, 29-30, 42-43, 45
Tokyo (Japan): 195-200
Tolten: 97, 232
tourist diver (defined): 208
tee shirt: *174*, 248
Tragedy of the Lusitania, The: 45
Treaty of Versailles: 48
Tritonia Corporation of Scotland: 59-61
Truman, President Harry: 131-132
Trzaska, Captain Mick: 282
Tucker, Teddy: 62-63
Turner, Captain William: 19, 23-24, 25-28, 30-35, *155*
Twilight Zone, The: 180
Tye, PFC: 189
U-20: 22-24, *155*
U-156: 263
U-853: *171*, 230-231, 270
U-boat: 16, 18-21, 24, 26, 30-32, 34-35, 38, 92, 99, 102, 103, 230, 263
Underwater Sports: 104
unionism: 292-296
United States: 29, 31, 36-37, 43, 58, 60, 65, 69
United States Coast Guard: 100, 244-247, 263, 278-279, 290
United States Navy: 65, 66, 100, 263
United States Navy Diving Manual: 247
United States Navy Standard Air Decompression Tables: 223, 262
University of Pennsylvania: 225-226, 279
unrestricted submarine warfare: 20-21, 38
Valaiti, Bruno: 109
Valley Forge General Hospital (Pennsylvania): 200-205
Vanderbilt, Alfred Gwynne: 49
Vanderbilt, Willliam H.: 50
van Name, Floyd: 282
Varanger: 97, 101, *171*

Verazanno Narrows (New York): 13
Versailles, Treaty of: 48
Vietnam: 125-144, 177-194, 220, 242, 254-255
Vietnam syndrome: 205
Viking Starlite: 91, 106-109
Virginia (state): 103
Virginia (wreck): see *Sommerstad*
War of 1812: 38
Warehouse, Kathy: 265-267
Washington, DC: 200
Washingtonian: 270
Washington's Crossing (Pennsylvania): 119
Watt, Captain James: 12
Waverly, Ed: 96
Weeks, Captain Teddy: 91
West, Frank: 109
Western World: 270
Westinghouse Lamp Company of New Jersey: 58
White Star: *174*, 263, 289
White Star line: 38, 42
Wilhelm Gustloff: 43
Williams, Master-at-Arms: 28
Williams, Palmer: 66-67
Wilson, Bob: 214, 217
Wilson, President Woodrow: 36-37, 44, 65
Winter Time Divers: 92
wireless: 11, 23, 28
Wizard of Oz: *172*
World Crisis, The: 40
World War One: 16-39, 40, 59, 263
World War Two: 43, 90, 103, 131, 200, 230, 236, 270
wreck-diver's code: 233-236, 274-275
Wreck of the Hesperus, The: 47
Wyatt, Jane: 219
Yasaka Maru: 53-54
Yawn, Corporal: 127-129, 143, 177-182, 184, 198, 207
Yearicks, Charlie: 112, 119-125
Young, James: 55-56
Young Men's Christian Association: 213-214
Young, Robert: 219
Ypres, Battles of: 40
Zuyder Zee: 57

The Lusitania Controversies

BOOK TWO:
Dangerous Descents into Shipwrecks and Law

(ISBN) 1-883056-07-1

Book Two begins where Book One ends: at the start of the 1980's - the decade of the *Doria*.

The *Andrea Doria* is synonymous with deep wreck-diving. Lying at a depth of 240 feet in unpredictable waters, it has attracted dedicated wreck-divers from around the world, and has been the site of some truly amazing feats and incredible penetrations into cold and utter darkness.

Elsewhere, the 1980's saw the expansion of shipwreck discovery and the exploration of wrecks that lie deeper than the *Doria*. Eventually, intrepid underwater explorers exceeded the productive depth limitation for breathing air. Nitrogen narcosis, lengthy decompression, and oxygen toxicity forced divers to experiment with alternative breathing mixtures that obviated the hazards induced by air at depth.

Blends of helium and oxygen extended the depth range beyond 300 feet. The true test occurred in 1990, with the author's successful dive to the German battleship *Ostfriesland*, at 380 feet. This pioneering venture and others that followed paved the way for the high tech, mixed-gas diving expedition to the *Lusitania* in 1994, in which the author was a participant.

Book Two describes all these expeditions in detail, as well as the territorial dispute over the *Lusitania's* legal status.

Available for $25 (postage paid) from:

Gary Gentile Productions
P.O. Box 57137
Philadelphia, PA 19111

http://www.pilot.infi.net/~boring/gentile.html

Books by the Author
Fiction

Vietnam	Science Fiction
Lonely Conflict	*Entropy*
	Return to Mars
Action/Adventure	*Silent Autumn*
Memory Lane	The Time Dragons Trilogy
Mind Set	*A Time for Dragons*
Supernatural	*Dragons Past*
The Lurking	*No Future for Dragons*

Nonfiction

Advanced Wreck Diving Guide	*Track of the Gray Wolf*
Ultimate Wreck Diving Guide	*Shipwrecks of New Jersey*

Available (postage paid) from: GARY GENTILE PRODUCTIONS
P.O. Box 57137
Nonfiction   Philadelphia, PA 19111

$25 *Andrea Doria: Dive to an Era* (hard cover)
$20 *USS San Diego: the Last Armored Cruiser*
$20 *Wreck Diving Adventures*
$20 *The Nautical Cyclopedia*
$20 *Primary Wreck Diving Guide*
$30 *The Technical Diving Handbook*
             Civil War ironclad **Monitor**
$25 (hard cover) *Ironclad Legacy: Battles of the USS Monitor*
$25 (video tape - VHS or PAL) *The Battle for the USS Monitor*
           The ***Lusitania*** Controversies (hard covers)
$25 Book One: *Atrocity of War and a Wreck-Diving History*
$25 Book Two: *Dangerous Descents into Shipwrecks and Law*
               The Popular Dive Guide Series
$20 *Shipwrecks of New York*
$20 *Shipwrecks of Delaware and Maryland*
$20 *Shipwrecks of Virginia*
$20 *Shipwrecks of North Carolina: from the Diamond Shoals North*
$20 *Shipwrecks of North Carolina: from Hatteras Inlet South*
             Wreck Diving Adventure Novel
$20 *The Peking Papers* (hard cover)

Website
http://www.pilot.infi.net/~boring/gentile.html